ANIMAL BEHAVIOR: AN INTRODUCTION

Edited by
Katyayani Tatiparti

Delve Publishing

Animal Behavior: An Introduction

Edited by Katyayani Tatiparti

Edition 2018

Delve Publishing

Published by

Delve Publishing LLC

708 3rd Avenue, 6th Floor

New York

NY 10801

United States of America

www.delvepublishing.com

ISBN: 978-1-68095-808-9 (Hardback)

Printed in United States of America

Library of Congress Control Number: 2016920691

About the Editor

Katyayani Tatiparti

Katyayani Tatiparti has a Master's degree in Pharmaceutical Sciences from JNTU, Hyderabad, India; with the topic of research for thesis being 'Self Micro Emulsifying Drug Delivery System for Carbamazepine (BCS Class II Drug)'. She also has a PG Diploma in Patents Law from NALSAR University, Hyderabad, India. She has been an IPR Professional and an Assistant Professor of Pharmaceutics. She is currently pursuing her second Master's degree in Pharmaceutical Sciences at Wayne State University, USA where she is pursuing research in cancer therapy.

Table of Contents

2 ANIMAL CONSCIOUSNESS

3 ANIMAL LANGUAGE

4 ELEPHANT COGNITION

5 EMOTION IN ANIMALS

6 NEUROETHOLOGY

7 OBSERVATIONAL LEARNING

8 PAIN IN ANIMALS

List of Abbreviations

AFCD	Agriculture, Fisheries and Conservation Department
AHA	American Humane Association
ACO	Ant Colony Optimization
BMPC	Beijing Municipal People's Congress
CNE	Computational Neuroethology
CF	Constant Frequency
ENR	Echo-to-noise ratio
EOD	Electric Organ Discharge
EEG	Electroencephalography
EA	Evolutionary Algorithms
FM	Frequency Modulated
fMRI	functional MRI
MRI	Magnetic Resonance Imaging
PSO	Particle Swarm Optimization
PTSD	Post-traumatic Stress Disorder
SAG	Screen Actors Guild
SPP	Self-propelled Particles
SDS	Stochastic Diffusion Search

Preface

The study of animal behavior begins with understanding how an animal's physiology and anatomy are integrated with its behavior. Both external and internal stimuli prompt behaviors – external information (e.g., threats from other animals, sounds, smells) or weather and internal information (e.g., hunger, fear). The study of behavior helps us to understand when and why animals, including humans, act. At one extreme, behavior is the link between the nervous system and the world outside an animal. At the other extreme, behavior is the way in which an animal establishes and maintains itself in its ecological niche. Behavior is a key link between an animal and its environment. Animals behavior includes communication, navigation, foraging, mating, parenting, and social cooperation. Each of these topics is a major subject within the study of behavior, and we will deal with each in a separate chapter. These issues all involve action and interaction; they are the externally expressed component of animal behavior. Behavior has internal underpinnings, as well. Motivation, learning, cognition and emotion shape an animal's behavioral decisions and therefore are critical in the study of behavior. Each of these topics has a neurobiological basis, so that their study links externally expressed behavior with the internal state of an animal. Scientists are drawn to the study of animal behavior for varied reasons and the field is extremely broad, ranging from research on feeding behavior and habitat selection to mating behavior and social organizations. Many scientists study animal behavior because it sheds light on human beings. Research on non-human primates, for instance, continues to offer valuable perspectives into the causes and evolution of individual, social, and reproductive human actions. Understanding why some animals help others at the potential cost of their own survival and reproduction, for example, not only gives us insight into their behavior but could also potentially help us to understand the underpinnings of our species' ideas of altruism and sacrifice.

Content Coverage

Chapters One and *Two* present an introduction to animal cognition and animal consciousness, respectively. Animal cognition describes the mental capacities of animals and the study of those capacities, animal consciousness is the quality or state of self-awareness within an animal, or of being aware of an external object or something within itself.

Chapters Three and *Four* are intended to focus on animal language and elephant cognition, separately. Animals communicate by using a variety of signs such as sounds or movements. Such signing may be considered complex enough to be called a form of language. Elephants manifest a wide variety of behaviors, including those associated with grief, learning, mimicry, play, altruism, use of tools, compassion, cooperation, self-awareness, memory, and communication.

Chapters Five and *Six* present a coverage of expression of emotions in animals and neuroethology, correspondingly. Emotions in animals are the subjective feelings and emotions experienced by non-human animals, whereas neuroethology deals with the study of the neural basis of natural behavior in animals.

Chapters Seven and *Eight* clarify about observational learning and pain in animals, separately. In animals, observational learning is often based on classical conditioning, in which an instinctive behavior is elicited by observing the behavior of another. Non-human animals cannot report their feelings to language-using humans in the same manner as human communication, but observation of their behavior provides a reasonable indication as to the extent of their pain.

CHAPTER 1

Animal Cognition

Chapter Contents

Animal cognition, or cognitive ethology, is the title given to a modern approach to the mental capacities of non human animals.

It has developed out of comparative psychology, but has also been strongly influenced by the approach of ethology, behavioral ecology, and evolutionary psychology.

Much of what used to be considered under the title of animal intelligence is now thought of under this heading.

Animal language acquisition, attempting to discern or understand the degree to which animal cognistics can be revealed by linguistics-related study, has been controversial among cognitive linguists.

INTRODUCTION

Experiments like the string-pulling task performed here by a Carib grackle provide insights into animal cognition.

Animal cognition describes the mental capacities of animals and the study of those capacities. The field developed from comparative psychology, including the study of animal conditioning and learning. It has also been strongly influenced by research in ethology, behavioral ecology, and evolutionary psychology, and hence the alternative name cognitive ethology is sometimes used. Many behaviors associated with the term *animal intelligence* are also subsumed within animal cognition.

A crab-eating macaque using a stone tool to crack open a nut

Researchers have examined animal cognition in mammals (especially primates, cetaceans, elephants, dogs, cats, horses, livestock, raccoons and rodents), birds (including parrots, corvids and pigeons), reptiles (lizards and snakes), fish and invertebrates (including cephalopods, spiders and insects).

HISTORICAL BACKGROUND

Morgan's Canon

Coined by 19th-century British psychologist C. Lloyd Morgan, Morgan's Canon remains a fundamental precept of comparative (animal) psychology. In its developed form, it states that:

In no case is an animal activity to be interpreted in terms of higher psychological processes if it can be fairly interpreted in terms of processes which stand lower in the scale of psychological evolution and development.

In other words, Morgan believed that anthropomorphic approaches to animal behavior were fallacious, and that people should only consider behaviour as, for example, rational, purposive or affectionate, if there is no other explanation in terms of the behaviours of more primitive life-forms to which we do not attribute those faculties.

From Anecdote to Laboratory

The behavior of non-human animals has captivated human imagination from antiquity, and over the centuries many writers have speculated about the animal mind, or its absence.

Speculation about animal intelligence gradually yielded to scientific study after Darwin placed humans and animals on a continuum, although Darwin's largely anecdotal approach to the topic would not pass scientific muster later on. Unsatisfied with the anecdotal method of Darwin and his protégé J. G. Romanes, E. L. Thorndike brought animal behavior into the laboratory for objective scrutiny. Thorndike's careful observations of the escape of cats, dogs, and chicks from puzzle boxes led him to conclude that what appears to the naive human observer to be intelligent behavior may be strictly attributable to simple associations. According to Thorndike, using Morgan's Canon, the inference of animal reason, insight, or consciousness is unnecessary and misleading. At about the same time, I. P. Pavlov began his seminal studies of conditioned reflexes in dogs. Pavlov quickly abandoned attempts to infer canine mental processes; such attempts, he said, led only to disagreement and confusion. He was, however, willing to propose unseen physiological processes that might explain his observations.

The Behavioristic half-Century

The work of Thorndike, Pavlov and a little later of the outspoken behaviorist John B. Watson set the direction of much research on animal behavior for more than half a century. During this time there was considerable progress in understanding simple associations; notably, around 1930 the differences between Thorndike's instrumental (or operant) conditioning and Pavlov's classical (or Pavlovian) conditioning were clarified, first by Miller and Kanorski, and then by B. F. Skinner.Many experiments on conditioning followed; they generated some complex theories, but they made little or no reference to intervening mental processes. Probably the most explicit dismissal of the idea that mental processes control behavior was the radical behaviorism of Skinner. This view seeks to explain behavior, including "private events" like mental images, solely by reference to the environmental contingencies impinging on the human or animal.

Despite the predominantly behaviorist orientation of research before 1960, the rejection of mental processes in animals was not universal during those years. Influential exceptions included, for example, Wolfgang Köhler and his insightful chimpanzees and Edward Tolman whose proposed cognitive map was a significant contribution to subsequent cognitive research in both humans and animals.

The Cognitive Revolution

Beginning around 1960, a "cognitive revolution" in research on humans gradually spurred a similar transformation of research with animals. Inference to processes not directly observable became acceptable and then commonplace. An important proponent of this shift in thinking was Donald O. Hebb, who argued that "mind" is simply a name for processes in the head that control complex behavior, and that it is both necessary and possible to infer those processes from behavior. Animals came to be seen as "goal seeking agents that acquire, store, retrieve, and internally process information at many levels of cognitive complexity.". The remainder of this article touches many areas of research that have appeared or greatly progressed since this seminal change in thinking, and many of the theoretical and empirical findings that have captured wide attention.

METHODS

The acceleration of research on animal cognition in the last 50 years or so has led to a rapid expansion in the variety of species studied and methods employed. The remarkable behavior of large-brained animals such as primates and cetacea has claimed special attention, but all sorts of mammals large and small, birds, fish, ants, bees, and others have been brought into the laboratory or observed in carefully controlled field studies. In the laboratory, animals push levers, pull strings, dig for food, swim in water mazes, or respond to images on computer screens in discrimination, attention, memory, and categorization experiments. Careful field studies explore memory for food caches, navigation by stars, communication, tool use, identification of conspecifics, and many other matters. Studies often focus on the behavior of animals in their natural environments and discuss the putative function of the behavior for the propagation and survival of the species. These developments reflect an increased cross-fertilization from related fields such as ethology and behavioral biology. Also, contributions from behavioral neuroscience are beginning to clarify the physiological substrate of some inferred mental process.

Some researchers have made effective use of a Piagetian methodology, taking tasks which human children are known to master at different stages of development, and investigating which of them can be performed by particular species. Others have been inspired by concerns for animal welfare and the management of domestic species: for example Temple Grandin has harnessed her unique expertise in animal welfare and the ethical treatment of farm livestock to highlight underlying similarities between humans and other animals. From a methodological point of view, one of the main risks in this sort of work is anthropomorphism, the tendency to interpret an animal's behavior in terms of human feelings, thoughts, and motivations.\

RESEARCH QUESTIONS

The common chimpanzee can use tools. This individual is using a stick to get food.

Human and animal cognition have much in common, and this is reflected in the research summarized below; most of the headings found here might also appear in an article on human cognition. Of course, research in the two also differs in important respects. Notably, much research with humans either studies or involves language, and much research with animals is related directly or indirectly to behaviors important to survival in natural settings. Following are summaries of some of the major areas of research in animal cognition.

Perception

Animals process information from eyes, ears, and other sensory organs to perceive the environment. Perceptual processes have been studied in many species, with results that are often similar to those in humans. Equally interesting are those perceptual processes that differ from, or go beyond those found in humans, such as echolocation in bats and dolphins, motion detection by skin receptors in fish, and extraordinary visual acuity, motion sensitivity and ability to see ultraviolet light in some birds.

Attention

Much of what is happening in the world at any moment is irrelevant to current behavior. Attention refers to mental processes that select relevant information, inhibit irrelevant information, and switch among these as the situation demands. Often the selective process is tuned before relevant information appears; such expectation makes for rapid selection of key stimuli when they become available. A large body of research has explored the way attention and expectation affect the behavior of non-human animals, and much of this work suggests that attention operates in birds, mammals and reptiles in much the same way that it does in humans.

Selective Learning

Animals trained to discriminate between two stimuli, say black versus white, can be said to attend to the "brightness dimension," but this says little about whether this dimension is selected in preference to others. More enlightenment comes from experiments that allow the animal to choose from several alternatives. For example, several studies have shown that performance is better on, for example, a color discrimination (e.g. blue vs green) after the animal has learned another color discrimination (e.g. red vs orange) than it is after training on a different dimension such as an X shape versus and O shape. The reverse effect happens after training on forms. Thus, the earlier learning appears to affect which dimension, color or form, the animal will attend to.

Other experiments have shown that after animals have learned to respond to one aspect of the environment responsiveness to other aspects is suppressed. In "blocking", for example, an animal is conditioned to respond to one stimulus ("A") by pairing that stimulus with reward or punishment. After the animal responds consistently to A, a second stimulus ("B") accompanies A on additional training trials. Later tests with the B stimulus alone elicit little response, suggesting that learning about B has been blocked by prior learning about A . This result supports the hypothesis that stimuli are neglected if they fail to provide new information. Thus, in the experiment just cited, the animal failed to attend to B because B added no information to that supplied by A. If true, this interpretation is an important insight into attentional processing, but this conclusion remains uncertain because blocking and several related phenomena can be explained by models of conditioning that do not invoke attention.

Divided Attention

Attention is a limited resource and is not an all-or-nothing response: the more attention devoted to one aspect of the environment, the less is available for others.A number of experiments have studied this in animals. In one experiment, a tone and a light are presented simultaneously to pigeons. The pigeons gain a reward only by choosing the correct combination of the two stimuli (e.g. a high frequency tone together with a yellow light). The birds perform well at this task, presumably by dividing attention between the two stimuli. When only one of the stimuli varies and the other is presented at its rewarded value, discrimination improves on the variable stimulus but discrimination on the alternative stimulus worsens. These outcomes are consistent with the notion that attention is a limited resource that can be more or less focused among incoming stimuli.

Visual search and Attentional Priming

As noted above, attention functions to select information that is of special use to the animal. Visual search typically calls for this sort of selection, and search tasks have been used extensively in both humans and animals to determine the characteristics of attentional selection and the factors that control it.

Experimental research on visual search in animals was initially prompted by field observations published by Luc Tinbergen (1960). Tinbergen observed that birds are selective when foraging for insects. For example, he found that birds tended to catch the same type of insect repeatedly even though several types were available. Tinbergen suggested that this prey selection was caused by an attentional bias that improved detection of one type of insect while suppressing detection of others. This "attentional priming" is commonly said to result from a pretrial activation of a mental representation of the attended object, which Tinbergen called a "searching image."

Tinbergen's field observations on priming have been supported by a number of experiments. For example, Pietrewicz and Kamil (1977, 1979) presented blue jays with pictures of tree trunks upon which rested either a moth of species A, a moth of species B, or no moth at all. The birds were rewarded for pecks at a picture showing a moth. Crucially, the probability with which a particular species of moth was detected was higher after repeated trials with that species (e.g. A, A, A,...) than it was after a mixture of trials (e.g. A, B, B, A, B, A, A...). These results suggest again that sequential encounters with an object can establish an attentional predisposition to see the object.

Another way to produce attentional priming in search is to provide an advance signal that is associated with the target. For example, if you hear a song sparrow you may be predisposed to detect a song sparrow in a shrub, or among other birds. A number of experiments have reproduced this effect in animal subjects.

Still other experiments have explored nature of stimulus factors that affect the speed and accuracy of visual search. For example, the time taken to find a single target increases as the number of items in the visual field increases. This rise in RT is steep if the distracters are similar to the target, less steep if they are dissimilar, and may not occur if the distracters are very different in from the target in form or color.

Concepts and Categories

Fundamental but difficult to define, the concept of "concept" was discussed for hundreds of years by philosophers before it became a focus of psychological study. Concepts enable humans and animals to organize the world into functional groups; the groups may be composed of perceptually similar objects or events, diverse things that have a common function, relationships such as same versus different, or relations among relations such as analogies. Extensive discussions on these matters together with many references may be found in Shettleworth (2010) Wasserman and Zentall (2006) and in Zentall *et al.* (2008). The latter is freely available online

Methods

Most work on animal concepts has been done with visual stimuli, which can easily be constructed and presented in great variety, but auditory and other stimuli have been used as well. Pigeons have been widely used, for they have excellent vision and are readily conditioned to respond to visual targets; other birds and a number of other animals have been studied as well. In a typical experiment, a bird or other animal confronts a computer monitor on which a large number of pictures appear one by one, and the subject gets a reward for pecking or touching a picture of a category item and no reward for non-category items. Alternatively, a subject may be offered a choice between two or more pictures. Many experiments end with the presentation of items never seen before; successful sorting of these items shows that the animal has not simply learned many specific stimulus-response associations. A related method, sometimes used to study relational concepts, is matching-to-sample. In this task an animal sees one stimulus and then chooses between two or more alternatives, one of which is the same as the first; the animal is then rewarded for choosing the matching stimulus.

Perceptual Categories

Perceptual categorization is said to occur when a person or animal responds in a similar way to a range of stimuli that share common features. For example, a squirrel climbs a tree when it sees Rex, Shep, or Trixie, which suggests that it categorizes all three as something to avoid. This sorting of instances into groups is crucial to survival. Among other things, an animal must categorize if it is to apply learning about one object (e.g. Rex bit me) to new instances of that category (dogs may bite).

Natural categorles

Many animals readily classify objects by perceived differences in form or color. For example, bees or pigeons quickly learn to choose any red object and reject any green object if red leads to reward and green does not. Seemingly much more difficult is an animal's ability to categorize natural objects that vary a great deal in color and form even while belonging to the same group. In a classic study, Richard J. Herrnstein trained pigeons to respond to the presence or absence of human beings in photographs. The birds readily learned to peck photos that contained partial or full views of humans and to avoid pecking photos with no human, despite great differences in the form, size, and color of both the humans displayed and in the non-human pictures. In follow-up studies, pigeons categorized other natural objects (e.g. trees) and after training they were able without reward to sort photos they had not seen before . Similar work has been done with natural auditory categories, for example, bird songs. Honeybees (*Apis mellifera*) are able to form concepts of "up" and "down".

Functional or Associative Categories

Perceptually unrelated stimuli may come to be responded to as members of a class if they have a common use or lead to common consequences. An oft-cited study by Vaughan (1988) provides an example. Vaughan divided a large set of unrelated pictures into two arbitrary sets, A and B. Pigeons got food for pecking at pictures in set A but not for pecks at pictures in set B. After they had learned this task fairly well, the outcome was reversed: items in set B led to food and items in set A did not. Then the outcome was reversed again, and then again, and so on. Vaughan found that after 20 or more reversals, associating reward with a few pictures in one set caused the birds to respond to the other pictures in that set without further reward, as if they were thinking "if these pictures in set A bring food, the others in set A must also bring food." That is, the birds now categorized the pictures in each set as functionally equivalent. Several other procedures have yielded similar results.

Relational or Abstract Categories

When tested in a simple stimulus matching-to-sample task (described above) many animals readily learn specific item combinations, such as "touch red if the sample is red, touch green if the sample is green." But this does not demonstrate that they distinguish between "same" and "different" as general concepts. Better evidence is provided if, after training, an animal successfully makes a choice that matches a novel sample that it has never seen before. Monkeys and chimpanzees do learn to do this, as do pigeons if they are given a great deal of practice with many different stimuli. However, because the sample is presented first, successful matching might mean that the animal is simply choosing the most recently seen "familiar" item rather than the conceptually "same" item. A number of studies have attempted to distinguish these possibilities, with mixed results.

Rule Learning

The use of rules has sometimes been considered an ability restricted to humans, but a number of experiments have shown evidence of simple rule learning in primates and also in other animals. Much of the evidence has come from studies of sequence learning in which the "rule" consists of the order in which a series of events occurs. Rule use is shown if the animal learns to discriminate different orders of events and transfers this discrimination to new events arranged in the same order. For example, Murphy *et al.* (2008) trained rats to discriminate between visual sequences. For one group ABA and BAB were rewarded, where A="bright light" and B="dim light." Other stimulus triplets were not rewarded. The rats learned the visual sequence, although both bright and dim lights were equally associated with reward. More importantly, in a second experiment with auditory stimuli, rats responded correctly to sequences of novel stimuli that were arranged in the same order as those previously learned. Similar sequence learning has been demonstrated in birds and other animals as well.

Memory

The categories that have been developed to analyze human memory (short term memory, long term memory, working memory) have been applied to the study of animal memory, and some of the phenomena characteristic of human short term memory (e.g. the serial position effect) have been detected in animals, particularly monkeys. However most progress has been made in the analysis of spatial memory; some of this work has sought to clarify the physiological basis of spatial memory and the role of the hippocampus; other work has explored the spatial memory of scatter-hoarder animals such as Clark's nutcracker, certain jays, tits and certain squirrels, whose ecological niches require them to remember the locations of thousands of caches, often following radical changes in the environment.

Memory has been widely investigated in foraging honeybees, *Apis mellifera*, which use both transient short-term working memory that is non-feeder specific and a feeder specific long-term reference memory. Memory induced in a free-flying honeybee by a single learning trial lasts for days and, by three learning trials, for a lifetime. Slugs, *Limax flavus*, have a short-term memory of approximately 1 min and long-term memory of 1 month.

Methods

As in humans, research with animals distinguishes between "working" or "short-term" memory from "reference" or long-term memory. Tests of working memory evaluate memory for events that happened in the recent past, usually within the last few seconds or minutes. Tests of reference memory evaluate memory for regularities such as "pressing a lever brings food" or "children give me peanuts."

Habituation

This is one of the simplest tests for memory spanning a short time interval. The test compares an animal's response to a stimulus or event on one occasion to its response on a previous occasion. If the second response differs consistently from the first, the animal must have remembered something about the first, unless some other factor such as motivation, sensory sensitivity, or the test stimulus has changed.

Delayed Response

Delayed response tasks are among the most useful methods used to study short-term memory in animals. Dating from research by Hunter (1913), the animal was shown a stimulus, such as a picture or a colored light, and a few seconds or minutes later the animal had to choose among alternative stimuli. In Hunter's studies, for example, a light appeared briefly in one of three goal boxes and then later the animal was allowed to choose among the boxes, finding food behind the one that had been lighted. Most research has been done with some variation of the "delayed matching-to-sample" task. For example, in the initial study with this task, a pigeon was presented with a flickering or steady light. Then, a few seconds later, two pecking keys were illuminated, one with a steady light and one with a flickering light. The bird got food if it pecked the key that matched the original stimulus.

A commonly-used variation of the matching-to-sample task requires the animal to use the initial stimulus to control a later choice between different stimuli. For example, if the initial stimulus is a black circle, the animal learns to choose "red" after the delay; if it is a black square, the correct choice is "green". Ingenious variations of this method have been used to explore many aspects of memory, including forgetting due to interference and memory for multiple items.

Radial arm maze

The radial arm maze is used to test memory for spatial location and to determine the mental processes by which location is determined. In a radial maze test, an animal is placed on a small platform from which paths lead in various directions to goal boxes; the animal finds food in one or more goal boxes. Having found food in a box, the animal must return to the central platform. The maze may be used to test both reference and working memory. Suppose, for example, that over a number of sessions the same 4 arms of an 8-arm maze always lead to food. If in a later test session the animal goes to a box that has never been baited, this indicates a failure of reference memory. On the other hand, if the animal goes to a box that it has already emptied during the same test session, this indicates a failure of working memory. Various confounding factors, such as odor cues, are carefully controlled in such experiments.

Water Maze

The water maze is used to test an animal's memory for spatial location and to discover how an animal is able to determine locations. Typically the maze is circular tank filled with water that has been made milky so that it is opaque. Located somewhere in the maze is small platform placed just below the surface of the water. When placed in the tank, the animal swims around until it finds and climbs up on the platform. With practice the animal finds the platform more and more quickly. Reference memory is assessed by removing the platform and observing the relative amount of time the animal spends swimming in the area where the platform had been located. Visual and other cues in and around the tank may be varied to assess the animal's reliance on landmarks and the geometric relations among them.

Spatial Cognition

Whether an animal ranges over a territory of measured in square kilometers or square meters, its survival typically depends on its ability to do such things as find a food source and then return to its nest. Sometimes such a task can be performed rather simply, for example by following a chemical trail. Typically, however, the animal must somehow acquire and use information about locations, directions, and distances. The following paragraphs outline some of the ways that animals do this.

- **Beacons** Animals often learn what their nest or other goal looks like, and if it is within sight they may simply move toward it; it is said to serve as a "beacon".
- **Landmarks** When an animal is unable to see its goal, it may learn the appearance of nearby objects and use these landmarks as guides. Researchers working with birds and bees have demonstrated this by moving prominent objects in the vicinity of nest sites, causing returning foragers to hunt for their nest in a new location.
- **Dead reckoning** Dead reckoning, also known as "path integration," is the process of computing one's position by starting from a known location and keeping track of the distances and directions subsequently traveled. Classic experiments have shown that the desert ant keeps track of its position in this way as it wanders for many meters searching for food. Though it travels in a randomly twisted path, it heads straight home when it finds food. However, if the ant is picked up and released some meters to the east, for example, it heads for a location displaced by the same amount to the east of its home nest.
- **Cognitive maps** Some animals appear to construct a cognitive map of their surroundings, meaning that they acquire and use information that enables them to compute how far and in what direction to go to get from one location to another. Such a map-like representation is thought to be used, for example, when an animal goes directly from one food source to another even though its previous experience has involved only travel between each source and home. Research in this area has also explored such topics as the use of geometric properties of the environment by rats and pigeons, and the ability of rats to

represent a spatial pattern in either radial arm mazes or water mazes. Spatial cognition is sometimes explored in visual search experiments in which a human or animal searches the environment for a particular object.

- **Detour behaviour** Some animals appear to have advanced understanding of their spatial environment and will not take the most direct route if this confers an advantage to them. Some jumping spiders take an indirect route to prey rather than the most direct route, thereby indicating flexibility in behaviour and route planning, and possibly insight learning.

Long-distance Navigation; Homing

Many animals travel hundreds or thousands of miles in seasonal migrations or returns to breeding grounds. They may be guided by the sun, the stars, the polarization of light, magnetic cues, olfactory cues, winds, or a combination of these.

It has been hypothesized that animals such as apes and wolves are good at spatial cognition because this skill is necessary for survival. Some researchers argue that this ability may have diminished somewhat in dogs because humans have provided necessities such as food and shelter during some 15,000 years of domestication.

Timing

Time of day: Circadian Rhythms

The behavior of most animals is synchronized with the earth's daily light-dark cycle. Thus, many animals are active during the day, others are active at night, still others near dawn and dusk. Though one might think that these "circadian rhythms" are controlled simply by the presence or absence of light, nearly every animal that has been studied has been shown to have a "biological clock" that yields cycles of activity even when the animal is in constant illumination or darkness.Circadian rhythms are so automatic and fundamental to living things — they occur even in plants - that they are usually discussed separately from cognitive processes, and the reader is referred to the main article (Circadian rhythms) for further information.

Interval Timing

Survival often depends on an animal's ability to time intervals. For example, rufous hummingbirds feed on the nectar of flowers, and they often return to the same flower, but only after the flower had had enough time to replenish its supply of nectar. In one experiment hummingbirds fed on artificial flowers that quickly emptied of nectar but were refilled at some fixed time (e.g. twenty minutes) later. The birds learned to come back to the flowers at about the right time, learning the refill rates of up to eight separate flowers and remembering how long ago they had visited each one.

The details of interval timing have been studied in a number of species. One of the most common methods is the "peak procedure". In a typical experiment, a rat in an operant chamber presses a lever for food. A light comes on, a lever-press brings a food pellet at a fixed later time, say 10 seconds, and then the light goes off. Timing is measured during occasional test trials on which no food is presented and the light stays on. On these test trials the rat presses the lever more and more until about 10 sec and then, when no food comes, gradually stops pressing. The time at which the rat presses most on these test trials is taken to be its estimate of the payoff time.

Experiments using the peak procedure and other methods have shown that animals can time short intervals quite exactly, can time more than one event at once, and can integrate time with spatial and other cues. Such tests have also been used for quantitative tests of theories of animal timing, such as Gibbon's Scalar Expectancy Theory ("SET"), Killeen's Behavioral Theory of Timing, and Machado's Learning to Time model. No one theory has yet gained unanimous agreement.

Tool and Weapon Use

Although tool use was long assumed to be a uniquely human trait, there is now much evidence that many animals use tools, including mammals, birds, fish, cephalopods and insects. Discussions of tool use often involve debate about what constitutes a "tool", and they often consider the relation of tool use to the animal's intelligence and brain size.

Mammals

Series of Photographs Showing a Bonobo Fishing for Termites.

Tool use has been reported many times in both wild and captive primates, particularly the great apes. The use of tools by primates is varied and includes hunting (mammals, invertebrates, fish), collecting honey, processing food (nuts, fruits, vegetables and seeds), collecting water, weapons and shelter. Research in 2007 shows that chimpanzees in the Fongolisavannah sharpen sticks to use as spears when hunting, considered the first evidence of systematic use of weapons in a species other than humans. Other mammals that spontaneously use tools in the wild or in captivity include elephants, bears, cetaceans, sea otters and mongooses.

A bonobo inserting a stick into a termite mound

The bonobo starts “fishing” for the termites

The bonobo withdraws the stick and begins eating the termites

The bonobo eats the termites extracted with the tool

Birds

Several species of birds have been observed to use tools in the wild, including warblers, parrots, Egyptian vultures, brown-headed nuthatches, gulls and owls. Some species, such as the woodpecker finch of the Galapagos Islands, use particular tools as an essential part of their foraging behavior. However, these behaviors are often quite inflexible and cannot be applied effectively in new situations. A great many species of birds build nests with a wide range of complexities, but although nest-building behaviour fulfills the criteria of some definitions of "tool-use", this is not the case with other definitions.

Several species of corvids have been trained to use tools in controlled experiments. One species examined extensively under laboratory conditions is the New Caledonian crow. One individual called "Betty" spontaneously made a wire tool to solve a novel problem. She was being tested to see whether she would select a wire hook rather than a straight wire to pull a little bucket of meat out of a well. Betty tried poking the straight wire at the meat. After a series of failures with this direct approach, she withdrew the wire and began directing it at the bottom of the well, which was secured to its base with duct tape. The wire soon became stuck, whereupon Betty pulled it sideways, bending it and unsticking it. She then inserted the hook into the well and extracted the meat. In all but one of 10 subsequent trials with only straight wire provided, she also made and used a hook in the same manner, but not before trying the straight wire first.

Fish

Several species of wrasses have been observed using rocks as anvils to crack bivalve (scallops, urchins and clams) shells. This behavior was first filmed in an orange-dotted tuskfish (*Choerodon anchorago*) in 2009 by Giacomo Bernardi. The fish fans sand to unearth the bivalve, takes it into its mouth, swims several meters to a rock, which it then uses as an anvil by smashing the mollusc apart with sideward thrashes of the head. This behaviour has also been recorded in a blackspot tuskfish (*Choerodon schoenleinii*) on Australia's Great Barrier Reef, yellowhead wrasse (*Halichoeres garnoti*) in Florida and a six-bar wrasse (*Thalassoma hardwicke*) in an aquarium setting. These species are at opposite ends of the phylogenetic tree in this family, so this behaviour may be a deep-seated trait in all wrasses.

Invertebrates

Some cephalopods are known to use coconut shells for protection or camouflage.

Ants of the species *Conomyrma bicolor* pick up stones and other small objects with their mandibles and drop them down the vertical entrances of rival colonies, allowing workers to forage for food without competition.

Reasoning and Problem Solving

It is clear that animals of quite a range of species are capable of solving problems that appear to require abstract reasoning;Wolfgang Köhler's (1917) work with chimpanzees is a famous early example. He observed that chimpanzees did not use trial and error to solve problems such as retrieving bananas hung out of reach. Instead, they behaved in a manner that was "unwaveringly purposeful," spontaneously placing boxes so that they could climb to reach the fruit. Modern research has identified similar behavior in animals usually thought of as much less intelligent, if appropriate pre-training is given.Causal reasoning has also been observed in rooks and New Caledonian crows.

It has been shown that Barbados bullfinches (*Loxigilla barbadensis*) from urbanized areas are better at innovative problem-solving tasks than bullfinches from rural environments, but that they did not differ in colour discrimination learning.

Cognitive Bias

Is the glass half empty or half full?

A **cognitive bias** refers to a systematic pattern of deviation from norm or rationality in judgment, whereby inferences about other individuals or situations may be drawn in an illogical fashion.

Cognitive bias is sometimes illustrated by using answers to the question "Is the glass half empty or half full?". Choosing "half empty" is supposed to indicate pessimism whereas choosing "half full" indicates optimism. To test this in animals, an individual is trained to anticipate that stimulus A, e.g. a 100 Hz tone, precedes a positive event, e.g. highly desired food is delivered when a lever is pressed by the animal. The same individual is trained to anticipate that stimulus B, e.g. a 900 Hz tone, precedes a negative event, e.g. bland food is delivered when the animal presses a lever. The animal is then tested by being given an intermediate stimulus C, e.g. a 500 Hz tone, and observing whether the animal presses the lever associated with the positive or negative reward. This has been suggested to indicate whether the animal is in a positive or negative mood.

In a study that used this approach, rats that were playfully tickled responded differently than rats that were simply handled. The rats that had been tickled were more optimistic than the handled rats. The authors suggested that they had demonstrated "...for the first time a link between the directly measured positive affective state and decision making under uncertainty in an animal model."

There is some evidence for cognitive bias in a number of species, including rats, dogs, rhesus macaques, sheep, chicks, starlings and honeybees.

Language

The modeling of human language in animals is known as animal language research. In addition to the ape-language experiments mentioned above, there have also been more or less successful attempts to teach language or language-like behavior to some non-primate species, including parrots and great spotted woodpeckers. Arguing from his own results with the animal Nim Chimpsky and his analysis of others results, Herbert Terrace criticized the idea that chimps can produce new sentences. Shortly thereafter Louis Herman published research on artificial language comprehension in the bottlenosed dolphin (Herman, Richards, & Wolz, 1984). Though this sort of research has been controversial, especially among cognitive linguists, many researchers agree that many animals can understand the meaning of individual words, and that some may understand simple sentences and syntactic variations, but there is little evidence that any animal can produce new strings of symbols that correspond to new sentences.

Insight

Wolfgang Köhler is usually credited with introducing the concept of insight into experimental psychology. Working with chimpanzees, Köhler came to dispute Edward Thorndike's theory that animals must solve problems gradually, by trial and error. He said that Thorndike's animals could only use trial and error because the situation precluded other forms of problem solving. He provided chimps with a relatively unstructured situation, and he observed sudden "ah-ha!" insightful changes of behavior, as, for example, when a chimp suddenly moved a box into position so that it could retrieve a banana. More recently, Asian elephants (*Elephas maximus*) were shown to exhibit similar insightful problem solving. A male was observed moving a box to a position where it could be stood upon to reach food that had been deliberately hung out of reach.

Numeracy

There a variety of studies indicate that animals are able to use and communicate quantitative information, and that some can count in a rudimentary way. Some examples of this research follow.

Elephants have been known to perform simple arithmetic, and rhesus monkeys and pigeons, in some sense, can count.

Ants are able to use quantitative values and transmit this information. For instance, ants of several species are able to estimate quite precisely numbers of encounters with members of other colonies on their feeding territories. Numeracy has been described in the yellow mealworm beetle (*Tenebrio molitor*) and the honeybee.

Western lowland gorillas given the choice between two food trays demonstrated the ability to choose the tray with more food items at a rate higher than chance after training. In a similar task, chimpanzees chose the option with larger amount of food. Salamanders given a choice between two displays with differing amounts of fruit flies, used as a food reward, reliably choose the display with more flies, as shown in a particular experiment.

Other experiments have been conducted that show animals' abilities to differentiate between non-food quantities. American black bears demonstrated quantity differentiation abilities in a task with a computer screen. The bears were trained to touch a computer monitor with a paw or nose to choose a quantity of dots in one of two boxes on the screen. Each bear was trained with reinforcement to pick a larger or smaller amount. During training, the bears were rewarded with food for a correct response. All bears performed better than what random error predicted on the trials with static, non-moving dots, indicating that they could differentiate between the two quantities. The bears choosing correctly in congruent (number of dots coincided with area of the dots) and incongruent (number of dots did not coincide with area of the dots) trials suggests that they were indeed choosing between quantities that appeared on the screen, not just a larger or smaller retinal image, which would indicate they are only judging size.

Bottlenose dolphins have shown the ability to choose an array with fewer dots compared to one with more dots. Experimenters set up two boards showing various numbers of dots in a poolside setup. The dolphins were initially trained to choose the board with the fewer number of dots. This was done by rewarding the dolphin when it chose the board with the fewer number of dots. In the experimental trials, two boards were set up, and the dolphin would emerge from the water and point to one board. The dolphins chose the arrays with fewer dots at a rate much larger than chance, indicating they can differentiate between quantities. A particular grey parrot, after training, has shown the ability to differentiate between the numbers zero through six using vocalizations. After number and vocalization training, this was done by asking the parrot how many objects there were in a display. The parrot was able to identify the correct amount at a rate higher than chance.Angelfish, when put in an unfamiliar environment will group together with conspecifics, an action named shoaling. Given the choice between two groups of differing size, the angelfish will choose the larger of the two groups. This can be seen with a discrimination ratio of 2:1 or greater, such that, as long as one group has at least twice the fish as another group, it will join the larger one.

Monitor lizards have been shown to be capable of numeracy, and some species can distinguish among numbers up to six.

Intelligence

Cognitive ability and intelligence in non-human animals cannot be measured with verbal scales, as in humans. Instead, intelligence in animals has been measured using a variety of methods that involve such things as habit reversal, social learning, and responses to novelty. However, as with humans, the intelligence attributed to an animal depends on the tasks used to measure that intelligence. In particular, an animal might appear "stupid" when measured with tasks appropriate for a human or an animal of another species, while that same animal might be a marvel of intelligence when given tasks like those it needs to survive in a natural setting.

Principal Component Analysis and factor analytic studies have shown that a single factor of intelligence is responsible for 47% of the individual variance in cognitive ability measures in primates and between 55% and 60% of the variance in mice. These values are similar to the accepted variance in IQ explained by a similar single factor known as the general factor of intelligence in humans (40-50%).

The general factor of intelligence, or *g* factor, is a psychometric construct that summarizes the correlations observed between an individual's scores on various measures of cognitive abilities. It has been suggested that *g* is related to evolutionary life histories and the evolution of intelligence as well as to social learning and cultural intelligence. Non-human models of *g* have been used in genetic and neurological research on intelligence to help understand the mechanisms behind variation in *g*.

Theory of Mind

Theory of mind is the ability to attribute mental states, e.g. intents, desires, pretending, knowledge, to oneself and others and to understand that others have desires, intentions, and perspectives that are different from one's own.

Some research with ravens provides an example of evidence for theory of mind in a non-human species. Ravens are members of the corvidae family, which is widely regarded as having high cognitive abilities. These birds have been observed to hide their food when dominant ravens are visible and audible at the same time. Based on this observation, ravens were tested for their understanding of "seeing" as a mental state. In a first step, the birds protected their cache when dominants were visible but not when they could only be heard from an adjacent room. In the next step, they had access to a small peephole which allowed them to see into the adjacent room. With the peephole open, the ravens guarded their caches against discovery when they could hear dominants in the adjacent room, even when the dominant's sounds were playbacks of recordings.

Consciousness

Mirror test with a baboon

The sense in which animals can be said to have consciousness or a self-concept has been hotly debated. The best known research technique in this area is the mirror test devised by Gordon G. Gallup, in which an animal's skin is marked in some way while it is asleep or sedated, and it is then allowed to see its reflection in a mirror; if the animal spontaneously directs grooming behavior towards the mark, that is taken as an indication that it is aware of itself. Self-awareness, by this criterion, has been reported for chimpanzees and also for other great apes, the European magpie, some cetaceans and a solitary elephant, but not for monkeys. The mirror test has been criticized researchers because it is entirely focused on vision, the primary sense in humans, while other species rely more heavily on other senses such as the olfactory sense in dogs.

It has been suggested that metacognition in some animals provides some evidence for cognitive self-awareness. The great apes, dolphins, and rhesus monkeys have demonstrated the ability to monitor their own mental states and use an "I don't know" response to avoid answering difficult questions. Unlike the mirror test, which reveals awareness of the condition of one's own body, this uncertainty monitoring is thought to reveal awareness of one's internal mental state. A 2007 study has provided some evidence for metacognition in rats, although this interpretation has been questioned.These species might also be aware of the strength of their memories.

Some researchers propose that an animal calls and other vocal behavior provide evidence as to consciousness. This idea arose from research on children's crib talk by Weir (1962) and in investigations of early speech in children by Greenfield and others (1976). Some such research has been done with a macaw (see Arielle).

In July, 2012 during the "Consciousness in Human and Nonhuman Animals" conference in Cambridge a group of scientists announced and signed a declaration with the following conclusions:

Convergent evidence indicates that non-human animals have the neuroanatomical, neurochemical, and neurophysiological substrates of conscious states along with the capacity to exhibit intentional behaviors. Consequently, the weight of evidence indicates that humans are not unique in possessing the neurological substrates that generate consciousness. Non-human animals, including all mammals and birds, and many other creatures, including octopuses, also possess these neurological substrates.

BIOLOGICAL CONSTRAINTS

Animals differ widely in many learning and cognitive tasks in ways that reflect their evolutionary history and their instinctual behaviors in natural environments. For example, dogs and rats easily learn to avoid an electric shock from the floor by moving to another part of the experimental chamber when they hear a tone preceding the shock; this is an appropriate response to a dangerous situation. However, hedgehogs fail to learn this avoidance behavior. This might seem to show the hedgehog's inability to learn, but the hedgehog's instinctive reaction to a threat is to curl up into a ball, a response that interferes with possible escape behavior in this situation.

Instinctive drift is another factor that can influence the interpretation of cognitive research. Instinctive drift is the tendency of an animal to revert to instinctive behaviors that can interfere with learned responses. The concept originated with Keller and Marian Breland when they taught a raccoon to put coins into a box. The raccoon drifted to its instinctive behavior of rubbing the coins with its paws, as it would do when foraging for food.

Hedgehogs instinctively roll into a ball when threatened, making them unsuitable for studies on aversion avoidance.

Animal ability to process and respond to stimuli is correlated with brain size. Small-brain animals tend to show simple behaviors that are less dependent on learning than those of large-brained animals. Vertebrates, particularly mammals, have large brains and complex behavior that changes with experience. A formula called the encephalization quotient (EC) expresses a relationship between brain and body size; it was developed by H.J. Jerison in the late 1960s. When the encephalization quotient is plotted as a curve, an animal with an EC above the curve is expected to show more cognitive ability than the average animal of its size, whereas an animal with an EC below the curve is expected to have less. Various formulas been suggested, but the equation $\boldsymbol{Ew(brain) = 0.12w(body)^{2/3}}$ has been found to fit data from a sample of mammals. The formula is suggestive at best, and should only be applied to non-mammals with extreme caution. For some of the other vertebrate classes, the power of 3/4 rather than 2/3 is sometimes used, and for many groups of invertebrates, the formula may not give meaningful results.

COGNITIVE FACULTY BY SPECIES

A traditionally common image is the *scala naturae*, the ladder of nature on which animals of different species occupy successively higher rungs, with humans typically at the top. However, rather than using such an arbitrary hierarchy, it seems more fruitful to understand cognitive capacities as adaptations to differing ecological niches.(see Shettleworth (1998), Reznikova (2007)).

Whether fairly or not, the performance of animals is often compared to that of humans on cognitive tasks. Not surprisingly, our closest biological relatives, the great apes, tend to perform most like humans. Among the birds, corvids and parrots have typically been found to perform well on human-like tasks. Some octopodes have also been shown to exhibit a number of higher-level skills such as tool use, but the amount of research on cephalopod intelligence is still limited.

Baboons have been shown to be capable of recognizing words.

REFERENCES

1. Shettleworth, S.J. (2010). Cognition, Evolution and Behavior (2 ed.). Oxford Press, New York.
2. Krueger, K.; Heinze, J. (2008). "Horse sense: social status of horses (Equus caballus) affects their likelihood of copying other horses` behavior" (PDF). Animal Cognition. 11 *(3): 431–439.* doi:10.1007/s10071-007-0133-0.
3. Krueger, K.; Farmer, K. & Heinze, J. (2013). "The effects of age, rank and neophobia on social learning in horses" (PDF). Animal Cognition. 17: 645–655. doi:10.1007/s10071-013-0696-x.
4. Fernandes, D.P.B.; da Silva, I.J.O.; Nazareno, A.C. & Donofre, A.C. (2015). "Farm animals's cognition and the tests used on its evaluation". J. Anim. Behav. Biometeorol. 3 *(1): 9–19. doi:10.14269/2318-1265/jabb.v3n1p9-19.*
5. Morgan, C.L. (1903). An Introduction to Comparative Psychology (2 ed.). W. Scott, London. p. 59.
6. Descartes, R. (1649), Passions of the Soul
7. Darwin, C. 1871, The descent of man, and selection in relation to sex
8. Romanes, J. G. 1883, Animal Intelligence
9. Thorndike, E. L. 1911, Animal intelligence.
10. Pavlov, I.P. 1928, Lectures on conditioned reflexes
11. Watson, J. B. (1913). "Psychology as the Behaviorist Views it.". *Psychological Review.* 20: 158–177. doi:10.1037/h0074428.
12. Miller, S.; Konorski, J. (1928). "Sur une forme particulière des reflexes conditionels". Comptes Rendus des Seances de la Societe de Biologie et de ses Filiales. 99: 1155–1157.
13. Skinner, B. F. (1932) The Behavior of Organisms
14. Hull, C. L. (1943) The Principles of Behavior
15. Skinner, B. F. About Behaviorism 1976
16. Köhler, W. (1917) The Mentality of Apes
17. Tolman, E. C. (1948) Cognitive maps in rats and men Psychological Review, 55, 189-208
18. Niesser, U. (1967) Cognitive Psychology
19. Hebb, D. O. 1958 A Textbook of Psychology
20. Menzel, R. & Fischer, J. (2010) Animal Thinking: Contemporary Issues in Comparative Cognition

21. Wasserman & Zentall (eds) (2006) Comparative Cognition

22. Grandin, Temple (2009) Animals Make Us Human: Creating the Best life for Animals (with Catherine Johnson)

23. Stebbins, W. C. & M. A. Berkley (1990) Comparative Perception,Vol. I, Basic Mechanisms; Vol. II, Complex Signals New York: Wiley.

24. Smith, E. E., and Kosslyn, S. M. (2007) "Cognitive Psychology: Mind and Brain" Pearson Prentice Hall.

25. Blough, D. S. (2006) Reaction-time explorations of visual attention, perception, and decision in pigeons. In E. A. Wasserman & T. R. Zentall (Eds) Comparative Cognition: Experimental Explorations of Animal Intelligence pp. 89-105. New York: Oxford.

26. N. J. Mackintosh (1983) Conditioning and Associative Learning New York: Oxford

27. Kamin, L. J. (1969) Predictability, surprise, attention, and conditioning. In Campbell and Church (eds.) Punishment and Aversive Behavior, New York: Appleton-Century-Crofts pp. 279-296

28. Mackintosh, N. J. (1994) Animal Learning and Cognition San Diego: Academic Press

29. Zentall, T. R. (2004). "Selective and divided attention in animals". Behavioural Processes. 69: 1–16. doi:10.1016/j.beproc.2005.01.004.

30. Blough, D. S. (1969). "Attention shifts in a maintained discrimination". Science. 166 *(3901): 125–126. Bibcode*:1969Sci...166..125B. doi:10.1126/science.166.3901.125.

31. Tinbergen, L (1960). "The natural control of insects in pine woods: I. Factors influencing the intensity of predation by songbirds". Archives Néerlandaises de Zoologie. 13: 265–343.

32. Pietrewicz, A. T.; Kamil, A. C. (1977). "Visual Detection of Cryptic Prey by Blue Jays (Cyanocitta cristata)". Science. 195 *(4278): 580–582. Bibcode*:1977Sci...195..580P. doi:10.1126/science.195.4278.580.

33. Pietrewicz, A. T. Kamil (1979). "Search image formation in the blue jay (Cyanocitta cristata).". *Science.* 204 *(4399): 1332–1333. Bibcode*:1979Sci...204.1332P. doi:10.1126/science.204.4399.1332.

34. Blough, P. M. (1989). "Attentional priming and visual search in pigeons". Journal of Experimental Psychology: Animal Learning and Cognition. 17: 292–298.

35. Kamil, A.C. & Bond, A. B. (2006) Selective attention, priming, and foraging behavior. In E. A. Wasserman and T. R. Zentall(eds) Comparative Cognition: Experimental Exploration of Animal Intelligence New York: Oxford

36. Blough, D. S. & Blough, P. M. (1990) Reaction-time assessments of visual processes in pigeons. In M. Berkley & W. Stebbins (Eds.) Comparative perception (pp. 245-276). New York:Wiley.

37. E. E. Smith & D. L. Medin (1981) Categories and Concepts Harvard Univ. Press

38. ^ Zentall, T. R.; Wasserman, E. A.; Lazareva, O. F.; Thompson, R. R. K.; Ratterman, M. J. (2008). "Concept Learning in Animals". Comparative Cognition & Behavior Reviews. 3: 13–45. doi:10.3819/ccbr.2008.30002.

39. Dooling, R. J., & Okanoya, K. (1995). Psychophysical methods for assessing perceptual categories. In G. M.Klump, R. J.Dooling, R. R.Fay, & W. C.Stebbins (Eds.)," Methods in Comparative Psychoacoustics" (pp. 307–318). Basel, Switzerland: Birkhäuser Verlag.

40. Herrnstein, R. J. (1964). "Complex Visual Concept in the Pigeon". Science. 146 *(3643): 549–551. Bibcode*:1964Sci...146..549H. doi:10.1126/science.146.3643.549.

41. Herrnstein, R. J. (1979). "Acquisition, Generalization, and Discrimination Reversal of a Natural Concept". J. of Experimental Psychology: Animal Behavior Processes. 5: 116–129. doi:10.1037/0097-7403.5.2.116.

42. R. S. Bhatt, E. A. Wasserman, W.F.J. Reynolds, & K. S.. Knauss (1988) "Conceptual behavior in pigeons: Categorization of both familiar and novel examples from four classes of natural and articifial stimuli." J. of Experimental Psychology: Animal Behavior Processes, 14, 219-234

43. Tu, H-W; Smith, E.; Dooling, R. J. (2011). "Acoustic and perceptual categories of vocal elements in the warble song of budgerigars (Melopsittacus undulates)". J. of Comparative Psychology. 125: 420–430. doi:10.1037/a0024396.

44. Avargues-Weber, A.; Dyer, A.G.; Giurfa, M. (2011). "Conceptualization of above and below relationships by an insect". Proceedings of The Royal Society B-Biological Sciences. 278: 898–905. doi:10.1098/rspb.2010.1891.

45. Vaughan, W.; Jr (1988). "Formation of equivalence sets in pigeons". Journal of Experimental Psychology: Animal Behavior Process. 14: 36–42. doi:10.1037/0097-7403.14.1.36.

46. See; D'Amato, M.; Columbo, M. (1988). "Representation of serial order in monkeys (Cebus apella")". Journal of Experimental Psychology: Animal Behavior Processes. 14: 11–139.

47. Murphy, R. A.; Mondragon, E.; Murphy, V. A. (2008). "Rule learning by rats"(PDF). Science. 319: 1849–1851. Bibcode:2008Sci...319.1849M. doi:10.1126/science.1151564.

48. Kundrey, S. M. A.; Strandell, B; Mathis, H.; Rowan, J. D. (2010). "Learning of monotonic and nonmonotonic sequences in domesticated horses (Equus callabus") and chickens (Gallus domesticus")". Learning and Motivation. 14: 213–223.

49. Wright, AA; Santiago, Hector C.; Sands, Stephen F.; Kendrick, Donald F.; Cook, Robert G. (1985). "Memory processing of serial lists by pigeons, monkeys, and people". Science. 229 *(4710): 287–289. Bibcode*:1985Sci...229..287W. doi:10.1126/science.9304205. PMID 9304205.

50. Balda, R.; Kamil, A. C. (1992). "Long-term spatial memory in Clark's nutcracker, Nucifraga columbiana". Animal Behaviour. 44 *(4): 761–769.* doi:10.1016/S0003-3472(05)80302-1.

51. Greggers, U.; Menzel, R. (1993). "Memory dynamics and foraging strategies of honeybees". Behavioral Ecology and Sociobiology. 32: 17–29. doi:10.1007/BF00172219.

52. Menzel, R. (1993). "Associative learning in honey-bees". Apidologie. 24 *(3): 157–168.* doi:10.1051/apido:19930301.

53. Wustenberg, D.; Gerber, B.; Menzel, R. (1998). "Long- but not medium-term retention of olfactory memory in honeybees is impaired by actinomycin D and anisomycin". European Journal of Neuroscience. 10: 2742–2745. doi:10.1046/j.1460-9568.1998.00319.x.

54. Hammer, M.; Menzel, R. (1995). "Learning and memory in the honeybee". Journal of Neuroscience. 15 *(3* Pt 1): 1617–1630. PMID 7891123.

55. Yamada, A.; Sekiguchi, T.; Suzuki, H.; Mizukami, A.; et al. (1992). "Behavioral analysis of internal memory states using cooling-induced retrograde anmesia in Limax flavus". The Journal of Neuroscience. 12 *(3): 729–735. PMID 1545237.*

56. Hunter, W. S. (1913) «The delayed reaction in animals and children» Behavior Monographs, 2

57. Blough, D. S. (1958). "Delayed matching in the pigeon". Journal of the Experimental Analysis of Behavior. 2: 151–160. doi:10.1901/jeab.1959.2-151.

58. Shettleworth, S. J. (2010) "Cognition, Evolution, and Behavior" New York: Oxford

59. Vorhees, C. V.; Williams, M. T. (2006). "Morris water maze: procedures for assessing spatial and related forms of learning and memory". Nature Protocols. 1: 848–858. doi:10.1038/nprot.2006.116.

60. ^ Jump up to:[a] [b] Animal Spatial Cognition:Comparative, Neural & Computational Approaches

61. Lund, Nick (2002). Animal cognition. Psychology Press. p. 4. ISBN 978-0-415-25298-0.

62. Sherwin, C.M., (2001). Can invertebrates suffer? Or, how robust is argument-by-analogy? Animal Welfare, 10 (supplement): S103-S118

63. Gauthreaux, Sidney A. (1980). Animal Migration, Orientation, and Navigation. Academic Press

64. Savolainen, Peter; Zhang, Ya-Ping; Luo, Jing; Lundeberg, Joakim; Leitner, Thomas (22 November 2002). "Genetic evidence for an East Asian origin of domestic dogs" (PDF). Science. 298 *(5598): 1060–1062. Bibcode*:2002Sci...298.1610S. doi:10.1126/science.1073906. PMID 12446907.

65. Fiset, Sylvain; Vickie Plourde (29 October 2012). "Object permanence in domestic dogs (Canis lupus familiaris) and gray wolves (Canis lupus)". Journal of Comparative Psychology. 127: 2. doi:10.1037/a0030595.

66. Brauer, Juliane; Juliane Kaminski; Julia Riedel; Josep Call; Michael Tomasello (2006). "Making inferences about the location of hidden food: social dog, causal ape". Journal of Comparative Psychology. 120 *(1): 38–47. doi:10.1037/0735-7036.120.1.38. PMID 16551163.*

67. Webb, Alex A.R. (2003). "The physiology of circadian rhythms in plants". New Phytologist. 160 *(2): 281–303. doi:10.1046/j.1469-8137.2003.00895.x.*

68. Henderson; Hurly, TA; Bateson, M; Healy, SD; et al. (2006). "Timing in free-living rufous hummingbirds, Selasphorus rufus". Current Biology. 16 *(5): 512–515. doi:10.1016/j.cub.2006.01.054. PMID 16527747.*

69. Gibbon, J (1977). "Scalar expectancy theory and Weber's law in animal timing". Psychological Review. 84 *(3): 279–325. doi:10.1037/0033-295x.84.3.279.*

70. Killeen, P. R. (1991). Behavior's time. In G. Bower (Ed.),The psychology of learning and motivation (Vol. 27, pp. 294–334). New York: Academic Press

71. Machado, A., & Pata, P. (2005). Testing the Scalar Expectancy Theory (SET) and the Learning to Time model (LeT) in a double bisection task. Behavior and Learning, 33, 111-122

72. Chimps Use "Spears" to Hunt Mammals, Study Says John Roach for National Geographic News (February 22, 2007) (accessed on June 12, 2010)

73. Hunt, G.R (1996). "Manufacture and use of hook-tools by New Caledonian crows". Nature. 379 *(6562): 249–251. Bibcode*:1996Natur.379..249H. doi:10.1038/379249a0.

74. Shettleworth, Sara J. (2012). "Do Animals Have Insight, and What Is Insight Anyway?" (PDF). Canadian Journal of Experimental Psychology. 66 *(4): 217–226.* doi:10.1037/a0030674.

75. Video shows first tool use by a fish

76. Bernardi, G. (2011). "The use of tools by wrasses (Labridae)." (PDF). doi:10.1007/s00338-011-0823-6. Retrieved July 7, 2013.

77. Finn, J. K.; Tregenza, T.; Tregenza, N. (2009). "Defensive tool use in a coconut-carrying octopus". Current Biology. 19 *(23): R1069–R1070. doi:10.1016/j.cub.2009.10.052. PMID 20064403.*

78. Michael H.J. Möglich; Gary D. Alpert (1979). "Stone dropping by Conomyrma bicolor (Hymenoptera: Formicidae): A new technique of interference competition". Behavioral Ecology and Sociobiology. 2 *(6): 105–113.* doi:10.1007/bf00292556. JSTOR 4599265.

79. For chimpanzees, see for example David Premack (1983) The Mind of an Ape

80. Wolfgang Köhler The Mentality of Apes (1917)

81. Pepperberg, I. M. (1999). The Alex Studies: Cognitive and Communicative Abilities of Grey Parrots. Cambridge MA: Harvard University Press.

82. Tebbich, Sabine; Seed, Amanda M.; Emery, Nathan J.; Clayton, Nicola S. (April 2007). "Non-tool-using rooks, Corvus frugilegus, solve the trap-tube problem". Anim Cogn. 10 *(2): 225–31.* doi:10.1007/s10071-006-0061-4. PMID 17171360.

83. Taylor, A.H; Hunt, G.R; Medina, F.S; Gray, R.D (January 2009). "Do New Caledonian crows solve physical problems through causal reasoning?". *Proc. R. Soc. B.* 276 *(1655): 247–254. doi:10.1098/rspb.2008.1107. PMC 2674354.* PMID 18796393.

84. Audet, J.N.; Ducatez, S. & Lefebvre, L. (2015). "The town bird and the country bird: problem solving and immunocompetence vary with urbanization". Behavioral Ecology. 27: 637–644. doi:10.1093/beheco/arv201.

85. Harding, EJ; Paul, ES; Mendl, M (2004). "Animal behaviour: cognitive bias and affective state.". *Nature.* 427 *(6972): 312. Bibcode*:2004Natur.427..312H. doi:10.1038/427312a. PMID 14737158.

86. Rygula, R; Pluta, H; P, Popik (2012). "laughing rats are optimistic". PLoS ONE. 7 *(12): e51959. Bibcode*:2012PLoSO...751959R. doi:10.1371/journal.pone.0051959. PMC 3530570. PMID 23300582.

87. Haselton, M. G.; Nettle, D. & Andrews, P. W. (2005). The evolution of cognitive bias. In D. M. Buss (Ed.), The Handbook of Evolutionary Psychology: Hoboken, NJ, US: John Wiley & Sons Inc. pp. 724–746.

88. Terrace, H., L.A. Petitto, R.J. Sanders, T.G. Bever(1979) Science 206 (4421): 891–902

89. Köhler, W. "Mentality of Apes," 1917.

90. Foerder, P.; Galloway, M.; Barthel, T.; Moore De, D. E.; Reiss, D. (2011). Samuel, Aravinthan, ed. "Insightful Problem Solving in an Asian Elephant". PLoS ONE. 6 *(8): e23251. Bibcode*:2011PLoSO...623251F. doi:10.1371/journal.pone.0023251. PMC 3158079. PMID 21876741.

91. Elephants show flair for arithmetic

92. Representation of the Numerosities 1-9 by Rhesus Macaques (Macaca) mulatto

93. Pigeons on Par with Primates in Numerical Competence

94. Reznikova, Zhanna; Ryabko, Boris (2001). "A Study of Ants' Numerical Competence". Electronic Transactions on Artificial Intelligence. 5: 111–126.

95. Reznikova, Zh. I. (2007). Animal Intelligence: From Individual to Social Cognition. Cambridge University Press

96. Reznikova, Zh. I. (1999). "Ethological mechanisms of population dynamic in species ant communities". Russian Journal of Ecology. 30 *(3): 187–197.*

97. Brown, M. J. F.; Gordon, D. M. (2000). "How resources and encounters affect the distribution of foraging activity in a seed-harvesting ants". Behavioral Ecology and Sociobiology. 47 *(3): 195–203.* doi:10.1007/s002650050011.

98. Carazo, P.; Font, E.; Forteza-Behrendt, E.; Desfilis, E.; et al. (2009). "Quantity discrimination in Tenebrio molitor: evidence of numerosity discrimination in an invertebrate?". *Animal Cognition.* 12 *(3): 463–470.* doi:10.1007/s10071-008-0207-7. PMID 19118405.

99. Dacke, M.; Srinivasan, M.V. (2008). "Evidence for counting in insects". Animal Cognition. 11 *(4): 683–689.* doi:10.1007/s10071-008-0159-y. PMID 18504627.

100. Anderson, U.S.; Stoinski, T.S.; Bloomsmith, M.A.; Marr, M.J.; Smith, A.D.; Maple, T.L. (2005). "Relative numerousness judgment and summation in young and old western lowland gorillas". Journal of Comparative Psychology. 119: 285–295. doi:10.1037/0735-7036.119.3.285.

101. Boysen, S.T.; Berntson, G.G.; Mukobi, K.L. (2001). "Size matters: impact of item size and quantity on array choice by chimpanzees (Pan troglodytes)". J. Comp. Psychol. 115: 106–110. doi:10.1037/0735-7036.115.1.106.

102. Uller, C.; Jaeger, R.; Guidry, G.; Martin, C. (2003). "Salamanders (Plethodon cinereus) go for more: rudiments of number in an amphibian". Anim Cogn. 6: 105–112. doi:10.1007/s10071-003-0167-x.

103. Vonk, J.; Beran, M.J. (2012). "Bears ‹count' too: quantity estimation and comparison in black bears, Ursus americanus". Animal Behaviour. 84 *(1): 231–238. doi:10.1016/j.anbehav.2012.05.001.*

104. Jaakkola, K.; Fellner, W.; Erb, L.; Rodriguez, M.; Guarino, E. (2005). "Understanding of the concept of numerically "less" by bottlenose dolphins (Tursiops truncatus)". Journal of Comparative Psychology. 119: 296–303. doi:10.1037/0735-7036.119.3.296.

105. Pepperberg, I (2006). "Grey parrot numerical competence: a review". Anim Cogn. 9: 377–391. doi:10.1007/s10071-006-0034-7.

106. Gómez-Laplaza, L.M.; Gerlai, R. (2010). "Can angelfish (Pterophyllum scalare) count? Discrimination between different shoal sizes follows Weber's law". Anim Cogn. 14: 1–9. doi:10.1007/s10071-010-0337-6.

107. King, Dennis & Green, Brian. 1999. Goannas: The Biology of Varanid Lizards. University of New South Wales Press. ISBN 0-86840-456-X, p. 43.

108. Reader, S. M.; Hager, Y.; Laland, K. N. (2011). "The evolution of primate general and cultural intelligence". Philosophical Transactions of the Royal Society B: Biological Sciences. 366 *(1567): 1017–1027. doi:10.1098/rstb.2010.0342.*

109. Locurto, C.; Durkin, E. "Problem-solving and individual differences in mice (Mus musculus) using water reinforcement". J Comp Psychol.

110. Locurto, C.; Scanlon, C. (1998). "Individual differences and a spatial learning factor in two strains of mice (Mus musculus)". J. Comp. Psychol. 112: 344–352. doi:10.1037/0735-7036.112.4.344.

111. Kamphaus, R. W. (2005). Clinical assessment of child and adolescent intelligence. Springer Science & Business Media.

112. Rushton, J. P. (2004). "Placing intelligence into an evolutionary framework or how g fits into the r–K matrix of life-history traits including longevity". Intelligence. 32 *(4): 321–328. doi:10.1016/j.intell.2004.06.003.*

113. Van Schaik, C. P.; Burkart, J. M. (2011). "Social learning and evolution: the cultural intelligence hypothesis". Philosophical Transactions of the Royal Society of London B: Biological Sciences. 366 *(1567): 1008–1016. doi:10.1098/rstb.2010.0304.*

114. Herrmann, E.; Call, J.; Hernández-Lloreda, M. V.; Hare, B.; Tomasello, M. (2007). "Humans have evolved specialized skills of social cognition: the cultural intelligence hypothesis". Science. 317 *(5843): 1360–1366. doi:10.1126/science.1146282.*

115. Plomin, R (2001). "The genetics of g in human and mouse". Nature Reviews Neuroscience. 2 *(2): 136–141.* doi:10.1038/35053584.

116. Anderson, B. (2000). The g factor in non-human animals. The nature of intelligence, (285), 79.

117. Premack, David; Woodruff, Guy (1978). "Does the chimpanzee have a theory of mind?". *Behavioral and Brain Sciences.* 1 *(4): 515–526.* doi:10.1017/S0140525X00076512.

118. Bugnyar, Thomas; Reber, Stephan A.; Buckner, Cameron (2016). "Ravens attribute visual access to unseen competitors". Nature Communications. 7: 10506. doi:10.1038/ncomms10506. PMC 4740864. PMID 26835849.

119. Helmut Prior; Ariane Schwarz; Onur Güntürkün (2008). De Waal, Frans, ed. "Mirror-Induced Behavior in the Magpie (Pica pica): Evidence of Self-Recognition" (PDF). PLoS Biology. Public Library of Science. 6 *(8): e202. doi:10.1371/journal.pbio.0060202. PMC 2517622.* PMID 18715117. Retrieved 2008-08-21.

120. Couchman, Justin J.; Coutinho, M. V. C.; Beran, M. J.; Smith, J. D. (2010). "Beyond Stimulus Cues and Reinforcement Signals: A New Approach to Animal Metacognition" (PDF). Journal of Comparative Psychology. 124 *(4)*: , 356 –368. doi:10.1037/a0020129. PMC 2991470. PMID 20836592.

121. Rats Capable Of Reflecting On Mental Processes

122. Foote, Allison L.; Crystal, J. D. (20 March 2007). "Metacognition in the Rat". Current Biology. 17 *(6): 551–555. doi:10.1016/j.cub.2007.01.061. PMC 1861845.* PMID 17346969.

123. Smith, J. David; Beran, M. J.; Couchman, J. J.; Coutinho, M. V. C. (2008). "The Comparative Study of Metacognition: Sharper Paradigms, Safer Inferences"(PDF). Psychonomic Bulletin & Review. 15 *(4): 679–691. doi:10.3758/PBR.15.4.679.*

124. Jozefowiez, J.; Staddon, J. E. R.; Cerutti, D. T. (2009). "Metacognition in animals: how do we know that they know?" (PDF). Comparative Cognition & Behavior Reviews. 4: 29–39. doi:10.3819/ccbr.2009.40003.

125. The Cambridge Declaration on Consciousness (PDF). Retrieved 12 August 2012.

126. Breland, K.; Breland, M. (1961). "The misbehavior of organisms". American Psychologist. 16: 681–684. doi:10.1037/h0040090.

127. Brett-Surman, Michael K.; Holtz, Thomas R.; Farlow, James O. (eds.). The complete dinosaur. Illustrated by Bob Walters (2nd ed.). Bloomington, Ind.: Indiana University Press. pp. 191–208. ISBN 978-0-253-00849-7.

128. Moore, J. (1999): Allometry, University of California, San Diego

129. Campbell, C.B.G.; Hodos, W. (1991). "The Scala Naturae revisited: Evolutionary scales and anagenesis in comparative psychology". J. Comp. Psychol. 105: 211–221. doi:10.1037/0735-7036.105.3.211.

130. Gopnik, Alison (May 2016). "How Animals Think; A new look at what humans can learn from nonhuman minds". The Atlantic. Retrieved 25 April 2016.

131. Finn, Julian K.; Tregenza, Tom; Norman, Mark D. (2009). "Defensive tool use in

a coconut-carrying octopus". Current Biology. 19 *(23): R1069–70. doi:10.1016/j. cub.2009.10.052. PMID 20064403.*

132. Baboons can learn to recognize words; Monkeys› ability suggests that reading taps into general systems of pattern recognition 12 April 2012 Nature
133. Baboons can recognize written words, study finds; The monkeys don›t assign meaning to them, but learn what letter combinations are common to real words, the study authors say April 12, 2012 Los Angeles Times
134. Baboons show their word skills; Reading may stem from a visual aptitude shared by all primates May 5, 2012

CHAPTER 2

Animal Consciousness

Chapter Contents

- **Introduction**
- **Philosophical Background**
- **Defining Consciousness**
- **Scientific Approaches**
- **Cambridge Declaration on Consciousness**
- **Examples**
- **Talking Animal**
- **Reported Cases by Species**

Elephant art has been a sensation for many years. Although the pachyderms are trained to paint -- how to hold a brush with their trunks, how to make strokes on a canvas -- from there, according to an interview with prominent elephant art academy founder Alex Melamid, trainers typically try to stay out of the elephants› artistic domain and let them work in their own individual styles.

Melamid said younger elephants are especially fond of painting. Some favor a speedy process, while others take a much different approach, painting in a methodical and unrushed manner, gently doling out drops of paint or pulling down graceful brushstrokes after appearing to carefully consider the progress of their creations.

INTRODUCTION

Animal consciousness, or animal awareness, is the quality or state of self-awareness within an animal, or of being aware of an external object or something within itself. In humans, consciousness has been defined as: sentience, awareness, subjectivity, qualia, the ability to experience or to feel, wakefulness, having a sense of selfhood, and the executive control system of the mind. Despite the difficulty in definition, many philosophers believe there is a broadly shared underlying intuition about what consciousness is.

According to the Cambridge Declaration on Consciousness, "near human-like levels of consciousness" have been observed in the African grey parrot.

The topic of animal consciousness is beset with a number of difficulties. It poses the problem of other minds in an especially severe form because animals, lacking the ability to use human language, cannot tell us about their experiences. Also, it is difficult to reason objectively about the question, because a denial that an animal is conscious is often taken to imply that it does not feel, its life has no value, and that harming it is not morally wrong. The 17th-century French philosopher René Descartes, for example, has sometimes been blamed for mistreatment of animals because he argued that only humans are conscious.

Philosophers who consider subjective experience the essence of consciousness also generally believe, as a correlate, that the existence and nature of animal consciousness can never rigorously be known. The American philosopher Thomas Nagel spelled out this point of view in an influential essay titled *What Is it Like to Be a Bat?*. He said that an organism is conscious "if and only if there is something that it is like to be that organism—something it is like *for* the organism"; and he argued that no matter how much we know about an animal's brain and behavior, we can never really put ourselves into the mind of the animal and experience its world in the way it does itself. Other thinkers, such as the cognitive scientist Douglas Hofstadter, dismiss this argument as incoherent. Several psychologists and ethologists have argued for the existence of animal consciousness by describing a range of behaviors that appear to show animals holding beliefs about things they cannot directly perceive—Donald Griffin›s 2001 book *Animal Minds* reviews a substantial portion of the evidence.

Animal consciousness has been actively researched for over 100 years. In 1927 the American functional psychologist Harvey Carr argued that any valid measure or understanding of awareness in animals depends on "an accurate and complete knowledge of its essential conditions in man". A more recent review concluded in 1985 that "the best approach is to use experiment (especially psychophysics) and observation to trace the dawning and ontogeny of self-consciousness, perception, communication, intention, beliefs, and reflection in normal human fetuses, infants, and children.

PHILOSOPHICAL BACKGROUND

René Descartes argued that only humans are conscious, and not other animals

The mind–body problem in philosophy examines the relationship between mind and matter, and in particular the relationship between consciousness and the brain. A variety of approaches have been proposed. Most are either dualist or monist. Dualism maintains a rigid distinction between the realms of mind and matter. Monism maintains that there is only one kind of stuff, and that mind and matter are both aspects of it. The problem was addressed by pre-Aristotelian philosophers,and was famously addressed by René Descartes in the 17th century, resulting in Cartesian dualism. Descartes believed that humans only, and not other animals have this non-physical mind.

The rejection of the mind–body dichotomy is found in French Structuralism, and is a position that generally characterized post-war French philosophy. The absence of an empirically identifiable meeting point between the non-physical mind and its physical extension has proven problematic to dualism and many modern philosophers of mind maintain that the mind is not something separate from the body. These approaches have been particularly influential in the sciences, particularly in the fields of sociobiology, computer science, evolutionary psychology, and the neurosciences.

Epiphenomenalism

Epiphenomenalism is the theory in philosophy of mind that mental phenomena are caused by physical processes in the brain or that both are effects of a common cause, as opposed to mental phenomena driving the physical mechanics of the brain. The impression that thoughts, feelings, or sensations cause physical effects, is therefore to be understood as illusory to some extent. For example, it is not the feeling of fear that produces an increase in heart beat, both are symptomatic of a common physiological origin, possibly in response to a legitimate external threat.

The history of epiphenomenalism goes back to the post-Cartesian attempt to solve the riddle of Cartesian dualism, i.e., of how mind and body could interact. La Mettrie, Leibniz and Spinoza all in their own way began this way of thinking. The idea that even if the animal were conscious nothing would be added to the production of behavior, even in animals of the human type, was first voiced by La Mettrie (1745), and then by Cabanis (1802), and was further explicated by Hodgson (1870) and Huxley (1874). Huxley (1874) likened mental phenomena to the whistle on a steam locomotive. However, epiphenomenalism flourished primarily as it found a niche among methodological or scientific behaviorism. In the early 1900s scientific behaviorists such as Ivan Pavlov, John B. Watson, and B. F. Skinner began the attempt to uncover laws describing the relationship between stimuli and responses, without reference to inner mental phenomena. Instead of adopting a form of eliminativism or mental fictionalism, positions that deny that inner mental phenomena exist, a behaviorist was able to adopt epiphenomenalism in order to allow for the existence of mind. However, by the 1960s, scientific behaviourism met substantial difficulties and eventually gave way to the cognitive revolution. Participants in that revolution, such as Jerry Fodor, reject epiphenomenalism and insist upon the efficacy of the mind. Fodor even speaks of epiphobia —fear that one is becoming an epiphenomenalist.

Thomas Henry Huxley defends in an essay titled *On the Hypothesis that Animals are Automata, and its History* an epiphenomenalist theory of consciousness according to which consciousness is a causally inert effect of neural activity— as the steam-whistle which accompanies the work of a locomotive engine is without influence upon its machinery . To this William James objects in his essay *Are We Automata?* by stating an evolutionary argument for mind-brain interaction implying that if the preservation and development of consciousness in the biological evolution is a result of natural selection, it is plausible that consciousness has not only been influenced by neural processes, but has had a survival value itself; and it could only have had this if it had been efficacious. Karl Popper develops in the book *The Self and Its Brain* a similar evolutionary argument.

Animal Ethics

Bernard Rollin of Colorado State University, the principal author of two U.S. federal laws regulating pain relief for animals, writes that researchers remained unsure into the 1980s as to whether animals experience pain, and veterinarians trained in the U.S. before 1989 were simply taught to ignore animal pain. In his interactions with scientists and other veterinarians, Rollin was regularly asked to prove animals are conscious and provide scientifically acceptable grounds for claiming they feel pain. Academic reviews of the topic are equivocal, noting that the argument that animals have at least simple conscious thoughts and feelings has strong support, but some critics continue to question how reliably animal mental states can be determined. A refereed journal *Animal Sentience*launched in 2015 by the Institute of Science and Policy of The Humane Society of the United States is devoted to research on this and related topics.

DEFINING CONSCIOUSNESS

"About forty meanings attributed to the term *consciousness* can be identified and categorized based on *functions* and *experiences*. The prospects for reaching any single, agreed-upon, theory-independent definition of consciousness appear remote."

Consciousness is an elusive concept that presents many difficulties when attempts are made to define it. Its study has progressively become an interdisciplinary challenge for numerous researchers, including ethologists, neurologists, cognitive neuroscientists, philosophers, psychologists and psychiatrists.

In 1976 Richard Dawkins wrote, "The evolution of the capacity to simulate seems to have culminated in subjective consciousness. Why this should have happened is, to me, the most profound mystery facing modern biology". In 2004, eight neuroscientists felt it was still too soon for a definition. They wrote an apology in "Human Brain Function":

We have no idea how consciousness emerges from the physical activity of the brain and we do not know whether consciousness can emerge from non-biological systems, such as computers... At this point the reader will expect to find a careful and precise definition of consciousness. You will be disappointed. Consciousness has not yet become a scientific term that can be defined in this way. Currently we all use the term *consciousness* in many different and often ambiguous ways. Precise definitions of different aspects of consciousness will emerge ... but to make precise definitions at this stage is premature."

Consciousness is sometimes defined as the quality or state of being aware of an external object or something within oneself. It has been defined somewhat vaguely as: subjectivity, awareness, sentience, the ability to experience or to feel, wakefulness, having a sense of selfhood, and the executive control system of the mind. Despite the difficulty in definition, many philosophers believe that there is a broadly shared underlying intuition about what consciousness is. Max Velmans and Susan Schneider wrote in *The Blackwell Companion to Consciousness*: "Anything that we are aware of at a given moment forms part of our consciousness, making conscious experience at once the most familiar and most mysterious aspect of our lives."

Related terms, also often used in vague or ambiguous ways, are:

- Awareness: the state or ability to perceive, to feel, or to be conscious of events, objects, or sensory patterns. In this level of consciousness, sense data can be confirmed by an observer without necessarily implying understanding. More broadly, it is the state or quality of being aware of something. In biological psychology, awareness is defined as a human›s or an animal›s perception and cognitive reaction to a condition or event.
- Self-awareness: the capacity for introspection and the ability to reconcile oneself as an individual separate from the environment and other individuals.
- Self-consciousness: an acute sense of self-awareness. It is a preoccupation with oneself, as opposed to the philosophical state of self-awareness, which is the awareness that one exists as an individual being; although some writers use both terms interchangeably or synonymously.
- Sentience: the ability to be aware (feel, perceive, or be conscious) of one›s surroundings or to have subjective experiences. Sentience is a minimalistic way

of defining consciousness, which is otherwise commonly used to collectively describe sentience plus other characteristics of the mind.

- Sapience: often defined as wisdom, or the ability of an organism or entity to act with appropriate judgment, a mental faculty which is a component of intelligence or alternatively may be considered an additional faculty, apart from intelligence, with its own properties.

Sentience (the ability to feel, perceive, or to experience subjectivity) is not the same as self-awareness (being aware of oneself as an individual). The mirror test is sometimes considered to be an operational test for self-awareness, and the handful of animals that have passed the mirror test are often considered to be self-aware. It remains debatable whether recognition of one's mirror image can be properly construed to imply full self-awareness, particularly given that robots are being constructed which appear to pass the test.

Much has been learned in neuroscience about correlations between brain activity and subjective, conscious experiences, and many suggest that neuroscience will ultimately explain consciousness; ...consciousness is a biological process that will eventually be explained in terms of molecular signaling pathways used by interacting populations of nerve cells.... However, this view has been criticized because consciousness has yet to be shown to be a process, and the so-called hard problem of relating consciousness directly to brain activity remains elusive.

SCIENTIFIC APPROACHES

Since Descartes's proposal of dualism, it became a general consensus that the mind had become a matter of philosophy and that science was not able to penetrate the issue of consciousness- that consciousness was outside of space and time. However, over the last 20 years, many scholars have begun to move toward a science of consciousness. Such notable neuroscientists that have led the move to neural correlates of the self and of consciousness are Antonio Damasio and Gerald Edelman. Damasio has demonstrated that emotions and their biological foundation play a critical role in high level cognition, and Edelman has created a framework for analyzing consciousness through a scientific outlook. The current problem consciousness researchers face involves explaining how and why consciousness arises from neural computation. In his research on this problem, Edelman has developed a theory of consciousness, in which he has coined the terms primary consciousness and secondary consciousness.

Eugene Linden, author of *The Parrot's Lament* suggests there are many examples of animal behavior and intelligence that surpass what people would suppose to be the boundary of animal consciousness. Linden contends that in many of these documented examples, a variety of animal species exhibits behavior that can only be attributed to emotion, and to a level of consciousness that we would normally ascribe only to our own species.

Philosopher Daniel Dennett counters that:

Consciousness requires a certain kind of informational organization that does not seem to be ‹hard-wired› in humans, but is instilled by human culture. Moreover, consciousness is not a black-or-white, all-or-nothing type of phenomenon, as is often assumed. The differences between humans and other species are so great that speculations about animal consciousness seem ungrounded. Many authors simply assume that an animal like a bat has a point of view, but there seems to be little interest in exploring the details involved.

Consciousness in mammals (this includes humans) is an aspect of the mind generally thought to comprise qualities such as subjectivity, sentience, and the ability to perceive the relationship between oneself and one's environment. It is a subject of much research in philosophy of mind, psychology, neuroscience, and cognitive science. Some philosophers divide consciousness into phenomenal consciousness, which is subjective experience itself, and access consciousness, which refers to the global availability of information to processing systems in the brain. Phenomenal consciousness has many different experienced qualities, often referred to as qualia. Phenomenal consciousness is usually consciousness *of* something or *about* something, a property known as intentionality in philosophy of mind.

Mirror test

Elephants can recognize themselves in a mirror.

The sense in which animals (or human infants) can be said to have consciousness or a self-concept has been hotly debated; it is often referred to as the debate over animal minds. The best known research technique in this area is the mirror test devised by Gordon G. Gallup, in which the skin of an animal (or human infant) is marked while it is asleep or sedated with a mark that cannot be seen directly but is visible in a mirror. The animal is then allowed to see its reflection in a mirror; if the animal spontaneously directs grooming behaviour towards the mark, that is taken as an indication that it is aware of itself. Over the past 30 years, many studies have found evidence that animals recognise themselves in mirrors. Self-awareness by this criterion has been reported for:

Apes

- common chimpanzees, bonobos, orangutans and gorillas.

Other land mammals

- elephants.

Cetaceans

- Bottlenose dolphin (*Tursiops truncatus*).
- Killer whales: Killer whales (*Orcinus orca*) and false killer whales (*Pseudorca crassidens*) may be able to recognize themselves in mirrors.

Birds

- magpies.

Until recently it was thought that self-recognition was absent from animals without a neocortex, and was restricted to mammals with large brains and well developed social cognition. However, in 2008 a study of self-recognition in corvids reported significant results for magpies. Mammals and birds inherited the same brain components from their last common ancestor nearly 300 million years ago, and have since independently evolved and formed significantly different brain types. The results of the mirror and mark tests showed that neocortex-less magpies are capable of understanding that a mirror image belongs to their own body. The findings show that magpies respond in the mirror and mark test in a manner similar to apes, dolphins and elephants. This is a remarkable capability that, although not fully concrete in its determination of self-recognition, is at least a prerequisite of self-recognition. This is not only of interest regarding the convergent evolution of social intelligence; it is also valuable for an understanding of the general principles that govern cognitive evolution and their underlying neural mechanisms. The magpies were chosen to study based on their empathy/lifestyle, a possible precursor for their ability of self-awareness. However even in the chimpanzee, the species most studied and with the most convincing findings, clear-cut evidence of self-recognition is not obtained in all individuals tested. Occurrence is about 75% in young adults and considerably less in young and old individuals. For monkeys, non-primate mammals, and in a number of bird species, exploration of the mirror and social displays were observed. However, hints at mirror-induced self-directed behavior have been obtained.

The mirror test has attracted controversy among some researchers because it is entirely focused on vision, the primary sense in humans, while other species rely more heavily on other senses such as the olfactory sense in dogs. A study in 2015 showed that the "sniff test of self-recognition (STSR)" provides evidence of self-awareness in dogs.

Language

Another approach to determine whether a non-human animal is conscious derives from passive speech research with a macaw (see Arielle). Some researchers propose that by passively listening to an animal's voluntary speech, it is possible to learn about the thoughts of another creature and to determine that the speaker is conscious. This type of research was originally used to investigate a child's crib speech by Weir (1962) and in investigations of early speech in children by Greenfield and others (1976).

Zipf›s law might be able to be used to indicate if a given dataset of animal communication indicate an intelligent natural language. Some researchers have used this algorithm to study bottlenose dolphin language (http://www.ee.columbia.edu/~xanadu/papers/suzuki05.pdf).

Pain or Suffering

Further arguments revolve around the ability of animals to feel pain or suffering. Suffering implies consciousness. If animals can be shown to suffer in a way similar or identical to humans, many of the arguments against human suffering could then, presumably, be extended to animals. Others have argued that pain can be demonstrated by adverse reactions to negative stimuli that are non-purposeful or even maladaptive. One such reaction is transmarginal inhibition, a phenomenon observed in humans and some animals akin to mental breakdown.

John Webster, a professor of animal husbandry at Bristol, argues:

People have assumed that intelligence is linked to the ability to suffer and that because animals have smaller brains they suffer less than humans. That is a pathetic piece of logic, sentient animals have the capacity to experience pleasure and are motivated to seek it, you only have to watch how cows and lambs both seek and enjoy pleasure when they lie with their heads raised to the sun on a perfect English summer›s day. Just like humans.

However, there is no agreement where the line between organisms that can feel pain and those that cannot should be drawn. Justin Leiber, a philosophy professor at Oxford University writes that:

Montaigne is ecumenical in this respect, claiming consciousness for spiders and ants, and even writing of our duties to trees and plants. Singer and Clarke agree in denying consciousness to sponges. Singer locates the distinction somewhere between the shrimp and the oyster. He, with rather considerable convenience for one who is thundering hard accusations at others, slides by the case of insects and spiders and bacteria, they pace Montaigne, apparently and rather conveniently do not feel pain. The intrepid Midgley, on the other hand, seems willing to speculate about the subjective experience of tapeworms ...Nagel ... appears to draw the line at flounders and wasps, though more recently he speaks of the inner life of cockroaches.

There are also some who reject the argument entirely, arguing that although suffering animals feel anguish, a suffering plant also struggles to stay alive (albeit in a less visible way). In fact, no living organism ‹wants› to die for another organism›s sustenance. In an article written for the New York Times, Carol Kaesuk Yoon argues that:

When a plant is wounded, its body immediately kicks into protection mode. It releases a bouquet of volatile chemicals, which in some cases have been shown to induce neighboring plants to pre-emptively step up their own chemical defenses and in other cases to lure in predators of the beasts that may be causing the damage to the plants. Inside the plant, repair systems are engaged and defenses are mounted, the molecular details of which scientists are still working out, but which involve signaling molecules coursing through the body to rally the cellular troops, even the enlisting of the genome itself, which begins churning out defense-related proteins ... If you think about it, though, why would we expect any organism to lie down and die for our dinner? Organisms have evolved to do everything in their power to avoid being extinguished. How long would any lineage be likely to last if its members effectively didn›t care if you killed them?

Cognitive Bias and Emotion

Cognitive bias in animals is a pattern of deviation in judgment, whereby inferences about other animals and situations may be drawn in an illogical fashion. Individuals create their own "subjective social reality" from their perception of the input. It refers to the question "Is the glass half empty or half full?", used as an indicator of optimism or pessimism. Cognitive biases have been shown in a wide range of species including rats, dogs, rhesus macaques, sheep, chicks, starlings and honeybees.

The neuroscientist Joseph LeDoux advocates avoiding terms derived from human subjective experience when discussing brain functions in animals. For example, the common practice of calling brain circuits that detect and respond to threats "fear circuits" implies that these circuits are responsible for feelings of fear. LeDoux argues that Pavlovian fear conditioning should be renamed Pavlovian threat conditioning to avoid the implication that fear is being acquired in rats or humans. Key to his theoretical change is the notion of survival functions mediated by survival circuits, the purpose of which is to keep organisms alive rather than to make emotions. For example, defensive survival circuits exist to detect and respond to threats. While all organisms can do this, only organisms that can be conscious of their own brain's activities can feel fear. Fear is a conscious experience and occurs the same way as any other kind of conscious experience: via cortical circuits that allow attention to certain forms of brain activity. LeDoux argues the only differences between an emotional and non-emotion state of consciousness are the underlying neural ingredients that contribute to the state.

Neuroscience

Neuroscience is the scientific study of the nervous system. It is a highly active interdisciplinary science that collaborates with many other fields. The scope of neuroscience has broadened recently to include molecular, cellular, developmental, structural, functional, evolutionary, computational, and medical aspects of the nervous system. Theoretical studies of neural networks are being complemented with techniques for imaging sensory and motor tasks in the brain. According to a 2008 paper, neuroscience explanations of psychological phenomena currently have a seductive allure , and seem to generate more public interest than explanations which do not contain neuroscientific information. They found that subjects who were not neuroscience experts judged that explanations with logically irrelevant neuroscience information were more satisfying than explanations without.

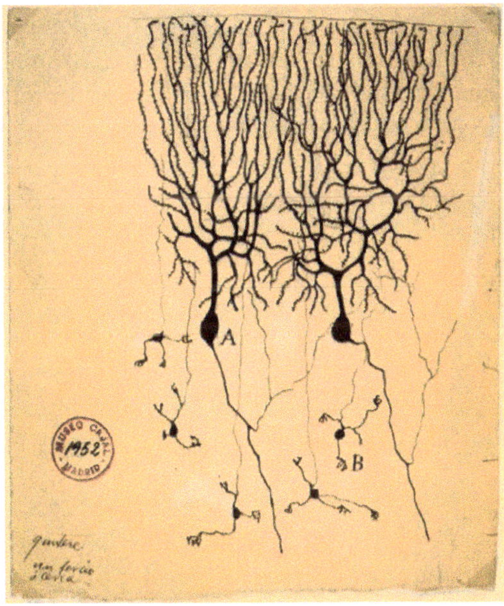

Drawing by Santiago Ramón y Cajal (1899) of neurons in the pigeon cerebellum

Neural correlates

The neural correlates of consciousness constitute the minimal set of neuronal events and mechanisms sufficient for a specific conscious percept. Neuroscientists use empirical approaches to discover neural correlates of subjective phenomena. The set should be *minimal* because, if the brain is sufficient to give rise to any given conscious experience, the question is which of its components is necessary to produce it.

Visual sense and representation was reviewed in 1998 by Francis Crick and Christof Koch. They concluded sensory neuroscience can be used as a bottom-up approach to studying consciousness, and suggested experiments to test various hypotheses in this research stream.

A feature that distinguishes humans from most animals is that we are not born with an extensive repertoire of behavioral programs that would enable us to survive on our own (physiological prematurity). To compensate for this, we have an unmatched ability to learn, i.e., to consciously acquire such programs by imitation or exploration. Once consciously acquired and sufficiently exercised, these programs can become automated to the extent that their execution happens beyond the realms of our awareness. Take, as an example, the incredible fine motor skills exerted in playing a Beethoven piano sonata or the sensorimotor coordination required to ride a motorcycle along a curvy mountain road. Such complex behaviors are possible only because a sufficient number of the subprograms involved can be executed with minimal or even suspended conscious control. In fact, the conscious system may actually interfere somewhat with these automated programs.

The growing ability of neuroscientists to manipulate neurons using methods from molecular biology in combination with optical tools depends on the simultaneous development of appropriate behavioural assays and model organisms amenable to large-scale genomic analysis and manipulation. A combination of such fine-grained neuronal analysis in animals with ever more sensitive psychophysical and brain imaging techniques in humans, complemented by the development of a robust theoretical predictive framework, will hopefully lead to a rational understanding of consciousness.

Neocortex

Previously researchers had thought that patterns of neural sleep exhibited by zebra finches needed a mammalian neocortex

The neocortex is a part of the brain of mammals. It consists of the grey matter, or neuronal cell bodies and unmyelinated fibers, surrounding the deeper white matter (myelinated axons) in the cerebrum. The neocortex is smooth in rodents and other small mammals, whereas in primates and other larger mammals it has deep grooves and wrinkles. These folds increase the surface area of the neocortex considerably without taking up too much more volume. Also, neurons within the same wrinkle have more opportunity for connectivity, while neurons in different wrinkles have less opportunity for connectivity, leading to compartmentalization of the cortex. The neocortex is divided into frontal, parietal, occipital, and temporal lobes, which perform different functions. For example, the occipital lobe contains the primary visual cortex, and the temporal lobe contains the primary auditory cortex. Further subdivisions or areas of neocortex are responsible for more specific cognitive processes. The neocortex is the newest part of the cerebral cortex to evolve (hence the prefix “neo”); the other parts of the cerebral cortex are the paleocortex and archicortex, collectively known as the allocortex. In humans, 90% of the cerebral cortex is neocortex.

Researchers have argued that consciousness arises in the neocortex, and therefore cannot arise in animals which lack a neocortex. For example, Rose argued in 2002 that the fishes have nervous systems that mediate effective escape and avoidance responses to noxious stimuli, but, these responses must occur without a concurrent, human-like awareness of pain, suffering or distress, which depend on separately evolved neocortex. Recently that view has been challenged, and many researchers now believe that animal consciousness can arise from homologous subcortical brain networks.

Attention

Attention is the cognitive process of selectively concentrating on one aspect of the environment while ignoring other things. Attention has also been referred to as the allocation of processing resources. Attention also has variations amongst cultures. Voluntary attention develops in specific cultural and institutional contexts through engagement in cultural activities with more competent community members.

Most experiments show that one neural correlate of attention is enhanced firing. If a neuron has a certain response to a stimulus when the animal is not attending to the stimulus, then when the animal does attend to the stimulus, the neuron's response will be enhanced even if the physical characteristics of the stimulus remain the same. In many cases attention produces changes in the EEG. Many animals, including humans, produce gamma waves (40–60 Hz) when focusing attention on a particular object or activity.

Extended Consciousness

Extended consciousness is an animal's autobiographical self-perception. It is thought to arise in the brains of animals which have a substantial capacity for memory and reason. It does not necessarily require language. The perception of a historic and future self arises from a stream of information from the immediate environment and from neural structures related to memory. The concept was popularised by Antonio Damasio and is used in biological psychology. Extended consciousness is said to arise in structures in the human brain described as *image spaces* and *dispositional spaces*. Image spaces imply areas where sensory impressions of all types are processed, including the focused awareness of the core consciousness. Dispositional spaces include convergence zones, which are networks in the brain where memories are processed and recalled, and where knowledge is merged with immediate experience.

Metacognition

Metacognition is defined as "cognition about cognition", or "knowing about knowing." It can take many forms; it includes knowledge about when and how to use particular strategies for learning or for problem solving. It has been suggested that metacognition in some animals provides evidence for cognitive self-awareness. The great apes, dolphins, and rhesus monkeys have demonstrated the ability to monitor their own mental states and use an "I don't know" response to avoid answering difficult questions. These species might also be aware of the strength of their memories. Unlike the mirror test, which relies primarily on body images and bodily self-awareness, uncertainty monitoring paradigms are focused on the kinds of mental states that might be linked to mental self-awareness.

There are generally two components of metacognition: knowledge about cognition, and regulation of cognition. Writings on metacognition can be traced back at least as far as *De Anima* and the *Parva Naturalia* of the Greek philosopher Aristotle. Metacognologists believe that the ability to consciously think about thinking is unique to sapient species and indeed is one of the definitions of sapience. There is evidence that rhesus monkeys and apes can make accurate judgments about the strengths of their memories of fact and monitor their own uncertainty, while attempts to demonstrate metacognition in birds have been inconclusive. A 2007 study has provided some evidence for metacognition in rats, but further analysis suggested that they may have been following simple operant conditioning principles, or a behavioral economic model.

Mirror Neurons

Mirror neurons are neurons that fire both when an animal acts and when the animal observes the same action performed by another. Thus, the neuron "mirrors" the behavior of the other, as though the observer were itself acting. Such neurons have been directly observed in primate and other species including birds. The function of the mirror system is a subject of much speculation. Many researchers in cognitive neuroscience and cognitive psychology consider that this system provides the physiological mechanism for the perception action coupling (see the common coding theory). They argue that mirror neurons may be important for understanding the actions of other people, and for learning new skills by imitation. Some researchers also speculate that mirror systems may simulate observed actions, and thus contribute to theory of mind skills, while others relate mirror neurons to language abilities. Neuroscientists such as Marco Iacoboni (UCLA) have argued that mirror neuron systems in the human brain help us understand the actions and intentions of other people. In a study published in March 2005 Iacoboni and his colleagues reported that mirror neurons could discern if another person who was picking up a cup of tea planned to drink from it or clear it from the table. In addition, Iacoboni and a number of other researchers have argued that mirror neurons are the neural basis of the human capacity for emotions such as empathy. Vilayanur S. Ramachandran has speculated that mirror neurons may provide the neurological basis of self-awareness.

Evolutionary Psychology

Consciousness is likely an evolved adaptation since it meets George Williams› criteria of species universality, complexity, and functionality, and it is a trait that apparently increases fitness. Opinions are divided as to where in biological evolution consciousness emerged and about whether or not consciousness has survival value. It has been argued that consciousness emerged (i) exclusively with the first humans, (ii) exclusively with the first mammals, (iii) independently in mammals and birds, or (iv) with the first reptiles.

Donald Griffin suggests in his book *Animal Minds* a gradual evolution of consciousness. Each of these scenarios raises the question of the possible survival value of consciousness.

In his paper Evolution of consciousness, John Eccles argues that special anatomical and physical adaptations of the mammalian cerebral cortex gave rise to consciousness. In contrast, others have argued that the recursive circuitry underwriting consciousness is much more primitive, having evolved initially in pre-mammalian species because it improves the capacity for interaction with both social *and* natural environments by providing an energy-saving "neutral" gear in an otherwise energy-expensive motor output machine. Once in place, this recursive circuitry may well have provided a basis for the subsequent development of many of the functions that consciousness facilitates in higher organisms, as outlined by Bernard J. Baars. Richard Dawkins suggested that humans evolved consciousness in order to make themselves the subjects of thought. Daniel Povinelli suggests that large, tree-climbing apes evolved consciousness to take into account one›s own mass when moving safely among tree branches. Consistent with this hypothesis, Gordon Gallup found that chimps and orangutans, but not little monkeys or terrestrial gorillas, demonstrated self-awareness in mirror tests.

The concept of consciousness can refer to voluntary action, awareness, or wakefulness. However, even voluntary behaviour involves unconscious mechanisms. Many cognitive processes take place in the cognitive unconscious, unavailable to conscious awareness. Some behaviours are conscious when learned but then become unconscious, seemingly automatic. Learning, especially implicitly learning a skill, can take place outside of consciousness. For example, plenty of people know how to turn right when they ride a bike, but very few can accurately explain how they actually do so.

Neural Darwinism

The American biologist Gerald Edelman distinguishes between primary and secondary consciousness:

- Primary consciousness: is the ability, found in humans and some animals, to integrate observed events with memory to create an awareness of the present and immediate past of the world around them. This form of consciousness is also sometimes called "sensory consciousness". Put another way, primary consciousness is the presence of various subjective sensory contents of consciousness such as sensations, perceptions, and mental images. For example, primary consciousness includes a person›s experience of the blueness of the ocean, a bird›s song, and the feeling of pain. Thus, primary consciousness refers to being mentally aware of things in the world in the present without any sense of past and future; it is composed of mental images bound to a time around the measurable present.
- Secondary consciousness: is an individual›s accessibility to their history and plans. The concept is also loosely and commonly associated with having awareness of one›s own consciousness. The ability allows its possessors to go beyond the limits of the remembered present of primary consciousness. The term was also coined by Gerald Edelman.

Primary consciousness can be defined as simple awareness that includes perception and emotion. As such, it is ascribed to most animals. By contrast, secondary consciousness depends on and includes such features as self-reflective awareness, abstract thinking, volition and metacognition.

Edelman›s theory focuses on two nervous system organizations: the brainstem and limbic systems on one side and the thalamus and cerebral cortex on the other side. The brain stem and limbic system take care of essential body functioning and survival, while the thalamocortical system receives signals from sensory receptors and sends out signals to voluntary muscles such as those of the arms and legs. The theory asserts that the connection of these two systems during evolution helped animals learn adaptive behaviors.

Other scientists have argued against Edelman›s theory, instead suggesting that primary consciousness might have emerged with the basic vegetative systems of the brain. That is, the evolutionary origin might have come from sensations and primal emotions arising from sensors and receptors, both internal and surface, signaling that the well-being of the creature was immediately threatened—for example, hunger for air, thirst, hunger, pain, and extreme temperature change. This is based on neurological data showing the thalamic, hippocampal, orbitofrontal, insula, and midbrain sites are the key to consciousness of thirst. These scientists also point out that the cortex might not be as important to primary consciousness as some neuroscientists have believed. Evidence of this lies in the fact that studies show that systematically disabling parts of the cortex in animals does not remove consciousness. Another study found that children born without a cortex are conscious. Instead of cortical mechanisms, these scientists emphasize brainstem mechanisms as essential to consciousness. Still, these scientists concede that higher order consciousness does involve the cortex and complex communication between different areas of the brain.

While animals with primary consciousness have long-term memory, they lack explicit narrative, and, at best, can only deal with the immediate scene in the remembered present. While they still have an advantage over animals lacking such ability, evolution has brought forth a growing complexity in consciousness, particularly in mammals. Animals with this complexity are said to have secondary consciousness. Secondary consciousness is seen in animals with semantic capabilities, such as the four great apes. It is present in its richest form in the human species, which is unique in possessing complex language made up of syntax and semantics. In considering how the neural mechanisms underlying primary consciousness arose and were maintained during evolution, it is proposed that at some time around the divergence of reptiles into mammals and then into birds, the embryological development of large numbers of new reciprocal connections allowed rich re-entrant activity to take place between the more posterior brain systems carrying out perceptual categorization and the more frontally located systems responsible for value-category memory. The ability of an animal to relate a present complex scene to its own previous history of learning conferred an adaptive evolutionary advantage. At much later evolutionary epochs, further re-entrant circuits appeared that linked semantic and linguistic performance to categorical and conceptual memory systems. This development enabled the emergence of secondary consciousness.

Ursula Voss of the Universität Bonn believes that the theory of protoconsciousness may serve as adequate explanation for self-recognition found in birds, as they would develop secondary consciousness during REM sleep. She added that many types of birds have very sophisticated language systems. Don Kuiken of the University of Alberta finds such research interesting as well as if we continue to study consciousness with animal models (with differing types of consciousness), we would be able to separate the different forms of reflectiveness found in today›s world.

For the advocates of the idea of a secondary consciousness, self-recognition serves as a critical component and a key defining measure. What is most interesting then, is the evolutionary appeal that arises with the concept of self-recognition. In non-human species and in children, the mirror test (see above) has been used as an indicator of self-awareness.

CAMBRIDGE DECLARATION ON CONSCIOUSNESS

Cambridge Declaration on Consciousness

The absence of a neocortex does not appear to preclude an organism from experiencing affective states. Convergent evidence indicates that non-human animals have the neuroanatomical, neurochemical, and neurophysiological substrates of conscious states along with the capacity to exhibit intentional behaviors. Consequently, the weight of evidence indicates that humans are not unique in possessing the neurological substrates that generate consciousness. Non-human animals, including all mammals and birds, and many other creatures, including octopuses, also possess these neurological substrates.:

In 2012, a group of neuroscientists attending a conference on "Consciousness in Human and non-Human Animals" at Cambridge University in the UK, signed *The Cambridge Declaration on Consciousness* (see box on the right).

In the accompanying text they unequivocally asserted:

- The field of Consciousness research is rapidly evolving. Abundant new techniques and strategies for human and non-human animal research have been developed. Consequently, more data is becoming readily available, and this calls for a periodic reevaluation of previously held preconceptions in this field. Studies of non-human animals have shown that homologous brain circuits correlated with conscious experience and perception can be selectively facilitated and disrupted to assess whether they are in fact necessary for those experiences. Moreover, in humans, new non-invasive techniques are readily available to survey the correlates of consciousness.
- The neural substrates of emotions do not appear to be confined to cortical structures. In fact, subcortical neural networks aroused during affective states in humans are also critically important for generating emotional behaviors in animals. Artificial arousal of the same brain regions generates corresponding behavior and feeling states in both humans and non-human animals. Wherever in the brain one evokes instinctual emotional behaviors in non-human animals, many of the ensuing behaviors are consistent with experienced feeling states, including those internal states that are rewarding and punishing. Deep brain stimulation of these systems in humans can also generate similar affective states. Systems associated with affect are concentrated in subcortical regions

where neural homologies abound. Young human and non-human animals without neocortices retain these brain-mind functions. Furthermore, neural circuits supporting behavioral/electrophysiological states of attentiveness, sleep and decision making appear to have arisen in evolution as early as the invertebrate radiation, being evident in insects and cephalopod mollusks (e.g., octopus)."

- Birds appear to offer, in their behavior, neurophysiology, and neuroanatomy a striking case of parallel evolution of consciousness. Evidence of near human-like levels of consciousness has been most dramatically observed in African grey parrots. Mammalian and avian emotional networks and cognitive microcircuitries appear to be far more homologous than previously thought. Moreover, certain species of birds have been found to exhibit neural sleep patterns similar to those of mammals, including REM sleep and, as was demonstrated in zebra finches, neurophysiological patterns previously thought to require a mammalian neocortex. Magpies in particular have been shown to exhibit striking similarities to humans, great apes, dolphins, and elephants in studies of mirror self-recognition.
- In humans, the effect of certain hallucinogens appears to be associated with a disruption in cortical feedforward and feedback processing. Pharmacological interventions in non-human animals with compounds known to affect conscious behavior in humans can lead to similar perturbations in behavior in non-human animals. In humans, there is evidence to suggest that awareness is correlated with cortical activity, which does not exclude possible contributions by subcortical or early cortical processing, as in visual awareness. Evidence that human and non-human animal emotional feelings arise from homologous subcortical brain networks provide compelling evidence for evolutionarily shared primal affective qualia.

EXAMPLES

A common image is the *scala naturae*, the ladder of nature on which animals of different species occupy successively higher rungs, with humans typically at the top. A more useful approach has been to recognize that different animals may have different kinds of cognitive processes, which are better understood in terms of the ways in which they are cognitively adapted to their different ecological niches, than by positing any kind of hierarchy.

Mammals

Dogs

Dogs were previously listed as non-self-aware animals. Because, traditionally, self-consciousness was evaluated via the mirror test, scientists can confirm that an animal species possess some sense of self. But dogs, as many other animals, are not visually oriented. In 2015 a study showed that the "sniff test of self-recognition (STSR)" provides significant evidences of self-awareness in dogs, and can play a crucial role in showing that this capacity is not a specific feature of only great apes, humans and a few other animals, but it depends on the way in which researchers try to verify it. According to the biologist Roberto Cazzolla Gatti, “the innovative approach to test the self-awareness with a smell test highlights the need to shift the paradigm of the anthropocentric idea of consciousness to a species-specific perspective”.

Birds

Alex engaged in numerical cognition

Research with captive African grey parrots, perhaps most notably Irene Pepperberg›s work with an individual named Alex, has demonstrated they possess the ability to associate simple human words with meanings, and to intelligently apply the abstract concepts of shape, colour, number, zero-sense, etc. According to Pepperberg and other scientists, they perform many cognitive tasks at the level of dolphins, chimpanzees, and even human toddlers. Another notable African grey is N›kisi, which in 2004 was said to have a vocabulary of over 950 words which she used in creative ways. For example, when Jane Goodall visited N›kisi in his New York home, he greeted her with Got a chimp? because he had seen pictures of her with chimpanzees in Africa.

The Eurasian magpie passes the mirror test

In 2011, research led by Dalila Bovet of Paris West University Nanterre La Défense, demonstrated African grey parrots were able to coordinate and collaborate with each other to an extent. They were able to solve problems such as two birds having to pull strings at the same time to obtain food. In another example, one bird stood on a perch to release a food-laden tray, while the other pulled the tray out from the test apparatus. Both would then feed. The birds were observed waiting for their partners to perform the necessary actions so their behaviour could be synchronized. The parrots appeared to express individual preferences as to which of the other test birds they would work with.

It was recently thought that self-recognition was restricted to mammals with large brains and highly evolved social cognition, but absent from animals without a neocortex. However, in 2008, an investigation of self-recognition in corvids was conducted revealing the ability of self-recognition in the magpie. Mammals and birds inherited the same brain components from their last common ancestor nearly 300 million years ago, and have since independently evolved and formed significantly different brain types. The results of the mirror test showed that although magpies do not have a neocortex, they are capable of understanding that a mirror image belongs to their own body. The findings show that magpies respond in the mirror test in a manner similar to apes, dolphins, killer whales, pigs and elephants. This is a remarkable capability that, although not fully concrete in its determination of self-recognition, is at least a prerequisite of self-recognition. This is not only of interest regarding the convergent evolution of social intelligence, it is also valuable for an understanding of the general principles that govern cognitive evolution and their underlying neural mechanisms. The magpies were chosen to study based on their empathy/ lifestyle, a possible precursor for their ability of self-awareness.

Invertebrates

Octopus travelling with shells collected for protection

Octopuses are highly intelligent, possibly more so than any other order of invertebrates. The level of their intelligence and learning capability are debated, but maze and problem-solving studies show they have both short- and long-term memory. Octopus have a highly complex nervous system, only part of which is localized in their brain. Two-thirds of an octopus› neurons are found in the nerve cords of their arms. Octopus arms show a variety of complex reflex actions that persist even when they have no input from the brain. Unlike vertebrates, the complex motor skills of octopuses are not organized in their brain using an internal somatotopic map of their body, instead using a non-somatotopic system unique to large-brained invertebrates. Some octopuses, such as the mimic octopus, move their arms in ways that emulate the shape and movements of other sea creatures.

In laboratory studies, octopuses can easily be trained to distinguish between different shapes and patterns. They reportedly use observational learning, although the validity of these findings is contested. Octopuses have also been observed to play: repeatedly releasing bottles or toys into a circular current in their aquariums and then catching them. Octopuses often escape from their aquarium and sometimes enter others. They have boarded fishing boats and opened holds to eat crabs.At least four specimens of the veined octopus (*Amphioctopus marginatus*) have been witnessed retrieving discarded coconut shells, manipulating them, and then reassembling them to use as shelter.

TALKING ANIMAL

A **talking animal** or **speaking animal** is any non-human animal that can produce sounds or gestures resembling those of a human language. Several species or groups of animals have developed forms of communication which superficially resemble verbal language, however, these are not defined as language because they lack one or more of the defining characteristics, i.e. grammar, syntax, recursion and displacement. Researchers have been successful in teaching some animals to make gestures similar to sign language. However, these animals fail to reach one or more of the criteria accepted as defining language.

On Imitation and Understanding

Clever Hans performing

The term refers to animals which can imitate (though not necessarily understand) human speech. Parrots, for example, repeat things nonsensically through exposure. It is an anthropomorphism to call this human speech, as it has no semantic grounding.

Clever Hans was a horse that was claimed to have been able to perform arithmetic and other intellectual tasks. After formal investigation in 1907, psychologist Oskar Pfungst demonstrated that the horse was not actually performing these mental tasks, but was watching the reaction of his human observers. The horse was responding to involuntary cues in the body language of his trainer or other people who asked the horse questions and knew the answer but were unaware they were providing such cues.

On Formality of Animal Language

A "formal language" requires a communication with a syntax as well as semantics. It is not sufficient for one to communicate information or even use symbols to communicate ideas. Researchers have attempted to teach great apes (gorillas, chimpanzees, and bonobos) spoken language with poor results, and sign language with significantly better results. However, even the best communicating great ape has shown an inability to grasp the idea of syntax and grammar, instead communicating at best at the same level as a pidgin language in humans. They are expressive and communicative, but lack the formality that remains unique to human speech.

Modern research shows that the key difference is the animal›s lack of asking questions and that formal syntax is merely a superficial detail, however Alex the parrot has been recorded as having asked an existential question. There are other differences as well, including poor precision, as shown by Kanzi the bonobo used the lexigram for *chase* interchangeably with that for *get*, though this behavior may not be the same for all animals. Research supports the idea that the linguistic limitations in animals are due to limited general brainpower (as opposed to lack of a specific module), and that words are created by breaking down sentences into pieces, making grammar more basic than semantics. The statement that syntax is the key difference between human and animal language is dubious.

REPORTED CASES BY SPECIES

Birds

- Research done by Dr. Irene Pepperberg indicates that parrots are capable of speaking in context and with intentional meaning. One of Pepperberg›s parrots, Alex, an African grey parrot, demonstrated the ability to assemble words out of letters.

Dogs

An owner hears a dog making a sound that resembles a phrase says the phrase back to the dog, who then repeats the sound and is rewarded with a treat. Eventually the dog learns a modified version of the original sound. Dogs have limited vocal imitation skills, so these sounds usually need to be shaped by selective attention and social reward.

- A dog on America's Funniest Home Videos named Fluffy, made noises that to some viewers resembled "I want my momma" after being asked "Do you want your momma? . Other videos showed other dogs making noises which to some viewers resembling Run around , I want it , I love momma and Hello .
- Odie, a pug who produced noises resembling "I love you" on demand, made appearances on several television shows.
- Paranormal researcher Charles Fort wrote in his book *Wild Talents* (1932) of several alleged cases of dogs that could speak English. Fort took the stories from contemporary newspaper accounts.
- A husky which produces vocalisations that to some viewers sound like ‹no› has appeared in the Daily Mail, the Mirror and the Huffington Post, amongst others.

Cats

- The case of a cat who was videotaped speaking purported human words and phrases such as "Oh my dog", "Oh Long John", "Oh Long Johnson", "Oh Don piano", "Why I eyes ya", and "All the live long day" became an Internet phenomenon in 2006. Footage of this cat, nicknamed Oh Long Johnson from one of the phrases spoken, was featured on *America›s Funniest Home Videos* in 1998, and a longer version of the clip (which revealed the animal was reacting to the presence of another cat) was aired in the UK. Clips from this video are prevalent on YouTube. The cat appeared as a character in Faith Hilling , the 226th episode of *South Park*, which aired on March 28, 2012.
- Miles v. City Council of Augusta, Georgia, in which the court found that the exhibition of a talking cat was considered an occupation for the purposes of municipal licensing law.

Others Hoover was a harbor seal who repeated common phrases heard around his exhibit at the New England Aquarium, including his name. He appeared in publications like *Reader's Digest* and *The New Yorker*, and television programs like *Good Morning America*.

- Gef the talking mongoose was an alleged talking animal who inhabited a small house on the Isle of Man, off the coast of Great Britain. Fringe authors believe Gef was a poltergeist, a strange animal or cryptid. Contemporary academics believe it was most likely a hoax.
- Batyr (1969–1993), an elephant from Kazakhstan, was reported to have a vocabulary of more than 20 phrases. Recordings of Batyr saying Batyr is good , Batyr is hungry , and words such as drink and give were played on Kazakh state radio in 1980.
- Kosik (born 1990) is an elephant able to imitate Korean words.

- Beluga whales can imitate the patterns of human speech. Noc, a captive beluga whale in the United States Navy›s Cold Ops program, could mimic some words well enough to confuse Navy divers on at least one occasion.
- It is not unusual for goats to make noises that sound like syllables from human words. Some videos of this behavior have ended up becoming popular on YouTube. An example from Tennessee, of a baby goat seeming to say "what what what? got over seven million views.

REFERENCES

1. The Cambridge Declaration on Consciousness (Archive) 7 July 2012. Written by Philip Low and edited by Jaak Panksepp, Diana Reiss, David Edelman, Bruno Van Swinderen, Philip Low and Christof Koch. University of Cambridge.
2. consciousness . Merriam-Webster. Retrieved 22 April 2014.
3. Robert van Gulick (2004). Consciousness . Stanford Encyclopedia of Philosophy.
4. Farthing G (1992). The Psychology of Consciousness. Prentice Hall. ISBN 978-0-13-728668-3.
5. John Searle (2005). Consciousness . In Honderich T. The Oxford companion to philosophy. Oxford University Press. ISBN 978-0-19-926479-7.
6. Colin Allen. Edward N. Zalta, ed. Animal consciousness . Stanford Encyclopedia of Philosophy (Summer 2011 Edition).
7. Peter Carruthers *(1999). "Sympathy and subjectivity". Australasian Journal of Philosophy. 77 (4): 465–482.* doi:10.1080/00048409912349231.
8. Thomas Nagel (1991). Ch. 12 What is it like to be a bat? . *Mortal Questions. Cambridge University Press. ISBN 978-0-521-40676-5.*
9. Douglas Hofstadter (1981). Reflections on What Is It Like to Be a Bat? . In Douglas Hofstadter; Daniel Dennett. The Mind›s I. Basic Books. pp. 403–414. ISBN 978-0-7108-0352-8.
10. Donald Griffin (2001). Animal Minds: Beyond Cognition to Consciousness. University of Chicago Press. ISBN 978-0-226-30865-4.
11. Burghardt, Gordon M (1985). Animal awareness: Current perceptions and historical perspective (PDF). American Psychologist. 40 *(8): 905–919. doi:10.1037/0003-066X.40.8.905. Archived from the original (PDF) on 16 September 2012.*
12. Carr, H (1927). The interpretation of the animal mind . Psychological Review. 34 *(2): 87*–106 . doi:10.1037/h0072244.
13. Robert M. Young (1996). The mind-body problem . In RC Olby; GN Cantor; JR Christie; MJS Hodges. Companion to the History of Modern Science (Paperback reprint of Routledge 1990 ed.). Taylor and Francis. pp. 702–11. ISBN 0415145783.
14. Robinson, Howard (3 November 2011). Edward N. Zalta, ed. Dualism . The Stanford Encyclopedia of Philosophy (Winter 2011 Edition).

15. Bryan S. Turner (2008). The Body and Society: Explorations in Social Theory(3rd ed.). Sage Publications. p. 78. ISBN 1412929873. ...a rejection of any dualism between mind and body, and a consequent insistence on the argument that the body is never simply a physical object but always an embodiment of consciousness.

16. Kim, Jaegwan (1995). Emergent properties . In Honderich, Ted. Problems in the Philosophy of Mind. Oxford Companion to Philosophy. Oxford: Oxford University Press. p. 240.

17. Pinel, J. (2009). Psychobiology (7th ed.). Pearson/Allyn and Bacon. ISBN 020554892X.

18. LeDoux, J. (2002). The Synaptic Self: How Our Brains Become Who We Are. Viking Penguin. ISBN 88-7078-795-8.

19. Russell, S.; Norvig, P. (2010). Artificial Intelligence: A Modern Approach (3rd ed.). Prentice Hall. ISBN 0136042597.

20. Dawkins, R. (2006). The Selfish Gene (3rd ed.). Oxford University Press. ISBN 0199291144.

21. Internet Encyclopedia of Philosophy. Epiphenomenalism . Retrieved 2012, Feb, 5. Check date values in: |access-date= (help)

22. Huxley, T. H. (1874). "On the Hypothesis that Animals are Automata, and its History , The Fortnightly Review, n.s.16:555-580. Reprinted in Method and Results: Essays by Thomas H. Huxley (New York: D. Appleton and Company, 1898)

23. Gallagher, S. 2006. "Where's the action?: Epiphenomenalism and the problem of free will . In W. Banks, S. Pockett, and S. Gallagher. Does Consciousness Cause Behavior? An Investigation of the Nature of Intuition (109-124). Cambridge, MA: MIT Press.

24. T. H. Huxley (1874). On the hypothesis that animals are automata, and its history . The Fortnightly Review. 16: 555–580.

25. W. James (1879). Are we automata? . *Mind.* 4: 1–22.

26. B. I. B. Lindahl (1997). Consciousness and biological evolution . Journal of Theoretical Biology. 187 *(4): 613–629. doi:10.1006/jtbi.1996.0394.*

27. Karl R. Popper; John C. Eccles *(1977). The Self and Its Brain. Springer International. ISBN 0-387-08307-3.*

28. Rollin, Bernard. The Unheeded Cry: Animal Consciousness, Animal Pain, and Science. New York: Oxford University Press, 1989, pp. xii, 117-118, cited in Carbone 2004, p. 150.

29. Griffin, DR; Speck, GB (2004). New evidence of animal consciousness (PDF). Animal Cognition. 7 *(1): 5–18.* doi:10.1007/s10071-003-0203-x. PMID 14658059.

30. The Ethics of research involving animals Nuffield Council on Bioethics, Accessed 27 February 2008 Archived 27 February 2008 at the Wayback Machine.

31. Allen, C (1998). Assessing animal cognition: ethological and philosophical perspectives (PDF). J. Anim. Sci. 76 *(1): 42–7. PMID 9464883.*

32. Animal Sentience An Interdisciplinary Journal on Animal Feeling

33. Vimal, RLP; Sansthana, DA (2010). On the Quest of Defining Consciousness (PDF). Mind and Matter. 8 *(1): 93–121.*

34. Hirstein, W (2013). Conscious states: Where are they in the brain and what are their necessary ingredients? . . Mens sana monographs. 11 *(1): 230. doi:10.4103/0973-1229.109343.*

35. Dulany, D. E. (2014). What Explains Consciousness? Or... What Consciousness Explains? . . Mens Sana Monographs. 12 *(1): 11. doi:10.4103/0973-1229.130283.*

36. De Sousa, Avinash (2013). Towards an integrative theory of consciousness: Part 1 (Neurobiological and cognitive models) . . Mens sana monographs. 11 *(1): 100. doi:10.4103/0973-1229.109335.*

37. Pereira Jr, A (2013). ›A Commentary On De Sousa›s Towards An Integrative Theory Of Consciousness › . Mens sana monographs. 11 *(1): 210. doi:10.4103/0973-1229.104495.*

38. Richard Dawkins, The Selfish Gene.

39. Human Brain Function, by Richard Frackowiak and 7 other neuroscientists, page 269 in chapter 16 The Neural Correlates of Consciousness (consisting of 32 pages), published 2004

40. consciousness . Merriam-Webster. Retrieved 4 June 2012.

41. Susan Schneider & Max Velmans *(2008). "Introduction". In Max* Velmans & Susan Schneider. The Blackwell Companion to Consciousness. Wiley. ISBN 978-0-470-75145-9.

42. Richard P. Lipka/Thomas M. Brinthaupt Self-perspectives Across the Life Span, p. 228, SUNY Press, 1992 ISBN 978-0-7914-1003-5

43. Gallup Jr GG; Anderson JR; Shillito DJ (2002). The mirror test . In Bekoff M; Allen C; Burghardt GM. The cognitive animal: Empirical and Theoretical perspectives on animal cognition (PDF). University of Chicago Press. pp. 325–333. Archived from the original (PDF) on 6 October 2014.

44. Haikonen POA (2007) Reflections of consciousness: The mirror test Proceedings of the 2007 AAAI Fall Symposium on Consciousness, pp. 67–71.

45. Asendorpf, J. B.; Warkentin, V.; Baudonniere, P.-M. (1996). Self-Awareness and Other Awareness II: Mirror Self-Recognition, Social Contingency Awareness, and Synchronic Imitation . Developmental Psychology. 32 *(2): 313–321. doi:10.1037/0012-1649.32.2.313.*

46. The World First Self-Aware Robot and the Success of Mirror Image Cognition(Lecture at the Karlsruhe University and the Munich University, Germany), 8 November 2006. Archived 3 February 2009 at the Wayback Machine.

47. Moreno, Raúl Arrabales (8 January 2008). Can a robot pass the mirror test? . *Conscious Robots.*

48. Eric R. Kandel (2007). In Search of Memory: The Emergence of a New Science of Mind. WW Norton. p. 9. ISBN 0393329372.

49. Oswald Hanfling (2002). Wittgenstein and the Human Form of Life. Psychology Press. pp. 108–109. ISBN 0415256453.

50. A term attributed to David Chalmers by Eugene O Mills (1999). Giving up on the hard problem of consciousness . In Jonathan Shear. Explaining Consciousness: The Hard Problem. MIT Press. p. 109. ISBN 026269221X.

51. Damasio, A. (1999). The Feeling of What Happens: Body, Emotion and the Making ofConsciousness. Harcourt Brace, 1999.

52. Damasio, A. (2003). Feelings of emotion and the self. In J. LeDoux, J. Debiec & H. Moss (Eds.), Self: from Soul to Brain (Vol. 1001, pp. 253-261). New York: New York Acad Sciences.

53. Chalmers D J, Facing up to the problem of consciousness, J Cons Stud, 2 (1995) 200.

54. Chalmers D J, The conscious mind: In search of a fundamental theory (Oxford University Press, New York) 1996.

55. Edelman, G. M. (2003). Naturalizing consciousness: a theoretical framework . Proc. Natl. Acad. Sci. U.S.A. 100 *(9): 5520–5524. doi:10.1073/pnas.0931349100.*

56. Edelman, G. M., Tononi, G. (2000). A Universe of Consciousness: How Matter Becomes Imagination. New York: Basic Books

57. Linden, Eugene (2000). The parrot›s lament: and other true tales of animal intrigue, intelligence, and ingenuity. Thorndike, Me.: G.K. Hall. ISBN 978-0-7838-9031-9.

58. Block, Ned (1995). On a confusion about a function of consciousness (PDF). The Behavioral and Brain Sciences. 18 *(2): 227–287.* doi:10.1017/s0140525x00038188.

59. Plotnik, JM; de Waal, FBM; Reiss, D (2006). Self-recognition in an Asian elephant . Proceedings of the National Academy of Sciences. 103 *(45): 17053–17057. Bibcode:*2006PNAS..10317053P. doi:10.1073/pnas.0608062103.

60. Bischof-Köhler, D. (1991). The development of empathy in infants. In M.E. Lamb & H. Keller (eds.), Infant Development. Perspectives from German speaking countries (245-273).

61. Prior, H; Schwarz, A; Gunturkun, O (2008). Mirror-induced behavior in the magpie (Pica pica): Evidence of self-recognition . PLOS Biology. 6 *(8): 1642–1650. doi:10.1371/journal.pbio.0060202. PMC 2517622.* PMID 18715117.

62. Gallup, GG Jr (1970). Chimpanzees: self-recognition . Science. 167 *(3914): 86–87. doi:10.1126/science.167.3914.86. PMID 4982211.*

63. Walraven, V; van Elsacker, L; Verheyen, R (1995). Reactions of a group of pygmy chimpanzees (Pan paniscus) to their mirror images: evidence of self-recognition . Primates. 36: 145–150. doi:10.1007/bf02381922.

64. Patterson FGP, Cohn RH (1994) Self-recognition and self-awareness in lowland gorillas. In: Parker ST, Mitchell RW, editors. Self-awareness in animals and humans: developmental perspectives. New York (New York): Cambridge University Press. pp. 273–290.

65. Marten, K.; Psarakos, S. (1995). Evidence of self-awareness in the bottlenose

dolphin (Tursiops truncatus) . In Parker, S.T.; Mitchell, R.; Boccia, M. Self-awareness in Animals and Humans: Developmental Perspectives. Cambridge University Press. pp. 361–379.

66. Reiss, D.; Marino, L. Mirror self-recognition in the bottlenose dolphin: A case of cognitive convergence . Proceedings of the National Academy of Sciences. 98 *(10): 5937–5942. doi:10.1073/pas.101086398 (inactive 2016-01-24).*

67. Delfour, F.; Marten, K. (2001). Mirror image processing in three marine mammal species: Killer whales (Orcinus orca), false killer whales (Pseudorca crassidens) and California sea lions (Zalophus californianus) . Behavioural Processes. 53 *(3): 181–190.* doi:10.1016/s0376-6357(01)00134-6. PMID 11334706.

68. Prior, Helmut; Schwarz, A; Güntürkün, O; De Waal, Frans (2008). De Waal, Frans, ed. Mirror-Induced Behavior in the Magpie (Pica pica): Evidence of Self-Recognition (PDF). PLoS Biology. Public Library of Science. 6 *(8): e202. doi:10.1371/journal.pbio.0060202. PMC 2517622. PMID 18715117. Retrieved 21 August 2008.*

69. Povinelli, DJ; Rulf, AB; Landau, KR; Bierschwale, DT (1993). Self-recognition in chimpanzees (Pan troglodytes): distribution, ontogeny, and patterns of emergence . J Comp Psychol. 107 *(4): 347–372. doi:10.1037/0735-7036.107.4.347. PMID 8112048.*

70. Hyatt, CW (1998). Responses of gibbons (Hylobates lar) to their mirror images . Am J Primatol. 45 *(3): 307–311.* doi:10.1002/(sici)1098-2345(1998)45:3<307::aid-ajp7>3.0.co;2-#.

71. Lea SEG (2010) Concept learning in nonprimate mammals: In search of evidence In D Mareschal, PC Quinn and SEG Lea, The Making of Human Concepts, pp. 173–199. Oxford University Press. ISBN 9780199549221.

72. The Superior Human? – Documentary. Transcription on the official website

73. Cazzolla Gatti, Roberto (2015). Self-consciousness: beyond the looking-glass and what dogs found there . Ethology Ecology & Evolution. 28 *(2): 232–240. doi:10.1080/03949370.2015.1102777. ISSN 0394-9370.*

74. Carter, Alan (2005). Animals, Pain and Morality . Journal of Applied Philosophy. 22 *(1): 17–22. doi:10.1111/j.1468-5930.2005.00289.x.*

75. The secret life of moody cows John Webster

76. Leiber, Justin (December 1988). ›Cartesian Linguistics?› . Philosophia. 18 *(4): 309–46.* doi:10.1007/BF02380646.

77. No Face, but Plants Like Life Too Carol Kaesuk Yoon, the New York Times

78. Haselton, M. G.; Nettle, D. & Andrews, P. W. (2005). The evolution of cognitive bias. . In D.M. Buss. The Handbook of Evolutionary Psychology. Hoboken, NJ, US: John Wiley & Sons Inc. pp. 724–746.

79. Bless, H.; Fiedler, K. & Strack, F. (2004). Social cognition: How individuals construct social reality. Hove and New York: Psychology Press. p. 2.

80. Mendl, M.; Burman, O.H.P.; Parker, R.M.A. & Paul, E.S. (2009). Cognitive

bias as an indicator of animal emotion and welfare: emerging evidence and underlying mechanisms . Applied Animal Behaviour Science. 118 *(3-4): 161–181. doi:10.1016/j.applanim.2009.02.023.*

81. Harding, EJ; Paul, ES; Mendl, M (2004). Animal behaviour: cognitive bias and affective state . Nature. 427 *(6972): 312.* doi:10.1038/427312a. PMID 14737158.

82. Rygula, R; Pluta, H; P, Popik (2012). Laughing rats are optimistic . PLoS ONE. 7 *(12): e51959. doi:10.1371/journal.pone.0051959. PMC 3530570. PMID 23300582.*

83. LeDoux, J (2012). Rethinking the emotional brain . Neuron. 73 *(4): 653–676. doi:10.1016/j.neuron.2012.02.004.*

84. LeDoux, JE (2014). Coming to terms with fear . Proc Natl Acad Sci USA. 111 *(8): 2871–2878. doi:10.1073/pnas.1400335111.*

85. LeDoux, JE (2015). Feelings: What are they & how does the brain make them? (PDF). Daedalus. 144 *(1): 96–111.* doi:10.1162/DAED_a_00319.

86. LeDoux JE (2015) Anxious: Using the Brain to Understand and Treat Fear and Anxiety, Penguin Publishing Group. ISBN 9781101619940.

87. Neuroscience . Merriam-Webster Medical Dictionary.

88. Weisberg, DS; Keil, FC; Goodstein, J; Rawson, E; Gray, JR (2008). The seductive allure of neuroscience explanations . Journal of Cognitive Neuroscience. 20 *(3): 470–477. doi:10.1162/jocn.2008.20040. PMC 2778755. PMID 18004955.*

89. Koch 2004, p. 304.

90. See here for a glossary of related terms.

91. Crick, F.; Koch, C. (1998). Consciousness and neuroscience. . Cereb Cortex. 8 *(2): 97–107. doi:10.1093/*cercor/8.2.97. PMID 9542889. Full text.

92. Beilock et al. 2002

93. Adamantidis A.R., Zhang F., Aravanis A.M., Deisseroth K. and de Lecea L. (2007) Neural substrates of awakening probed with optogenetic control of hypocretin neurons . Nature.

94. Rose, James D (2002). The neurobehavioral nature of fishes and the question of awareness and pain (PDF). Reviews in Fisheries Science. 10 *(1): 1–38.* doi:10.1080/20026491051668. Archived from the original (PDF) on 10 October 2012.

95. Anderson, John R. (2004). Cognitive psychology and its implications (6th ed.). Worth Publishers. p. 519. ISBN 978-0-7167-0110-1.

96. Chavajay, Pablo; Barbara Rogoff (1999). Cultural variation in management of attention by children and their caregivers . Developmental Psychology. 35 *(4): 1079–1090. doi:10.1037/0012-1649.35.4.1079. PMID 10442876.*

97. Kaiser J, Lutzenberger W (2003). Induced gamma-band activity and human brain function . Neuroscientist. 9 *(6): 475–84.* doi:10.1177/1073858403259137. PMID 14678580.

98. 'Layers' or 'levels' of consciousness?

99. Time Magazine 1999 'Mystery Of Consciousness'

100. Metcalfe, J., & Shimamura, A. P. (1994). Metacognition: knowing about knowing. Cambridge, MA: MIT Press.

101. Couchman, Justin J.; Coutinho, M. V. C.; Beran, M. J.; Smith, J. D. (2010). Beyond Stimulus Cues and Reinforcement Signals: A New Approach to Animal Metacognition (PDF). Journal of Comparative Psychology. 124 *(4): 356–368.* doi:10.1037/a0020129. PMC 2991470. PMID 20836592. Archived from the original (PDF) on 24 March 2012.

102. Schraw, Gregory (1998). Promoting general metacognitive awareness . Instructional Science. 26 *(1/2): 113–125.* doi:10.1023/A:1003044231033.

103. Oxford Psychology Dictionary;metacognition

104. Couchman, Justin J.; Coutinho, M. V. C.; Beran, M. J.; Smith, J. D. (2010). Beyond Stimulus Cues and Reinforcement Signals: A New Approach to Animal Metacognition (PDF). Journal of Comparative Psychology. 124 *(4): 356–368.* doi:10.1037/a0020129. PMC 2991470. PMID 20836592.

105. Metacognition: Known unknowns. Issue 2582 of New Scientist magazine, subscribers only.

106. Rats Capable Of Reflecting On Mental Processes

107. Foote, AL; Crystal, JD (March 2007). Metacognition in the rat . Curr. Biol. 17 *(6): 551–5. doi:10.1016/j.cub.2007.01.061. PMC 1861845. PMID 17346969.*

108. Foote, Allison L.; Crystal, J. D. (20 March 2007). Metacognition in the Rat . Current Biology. 17 *(6): 551–555. doi:10.1016/j.cub.2007.01.061. PMC 1861845. PMID 17346969.*

109. Smith, J. David; Beran, M. J.; Couchman, J. J.; Coutinho, M. V. C. (2008). The Comparative Study of Metacognition: Sharper Paradigms, Safer Inferences (PDF). Psychonomic Bulletin & Review. 15 *(4): 679–691. doi:10.3758/pbr.15.4.679.*

110. Jozefowiez, J.; Staddon, J. E. R.; Cerutti, D. T. (2009). Metacognition in animals: how do we know that they know? (PDF). Comparative Cognition & Behavior Reviews. 4: 29–39. doi:10.3819/ccbr.2009.40003.

111. Rizzolatti, Giacomo; Craighero, Laila (2004). The mirror-neuron system (PDF). Annual Review of Neuroscience. 27: 169–192. doi:10.1146/annurev.neuro.27.070203.144230. PMID 15217330.

112. Keysers, Christian (2010). Mirror Neurons (PDF). Current Biology. 19 *(21): R971–973. doi:10.1016/j.cub.2009.08.026. PMID 19922849.*

113. Keysers, Christian (23 June 2011). The Empathic Brain. Kindle.

114. Christian Keysers; Valeria Gazzola (2006). Anders; Ende; Unghofer; Kissler; Wildgruber, eds. Towards a unifying neural theory of social cognition (PDF). Progress in Brain Research. 156: 379–401. doi:10.1016/S0079-6123(06)56021-2. ISSN 0079-6123. Archived from the original (PDF) on 30 June 2007.

115. Michael Arbib, The Mirror System Hypothesis. Linking Language to Theory of Mind, 2005, retrieved 17 February 2006

116. Théoret, Hugo; Pascual-Leone, Alvaro (2002). Language Acquisition: Do as You Hear . Current Biology. 12 *(21): R736–7.* doi:10.1016/S0960-9822(02)01251-4. PMID 12419204.

117. Blakeslee, Sandra (10 January 2006). Cells That Read Minds . New York Times. Science.

118. Oberman, L.; Ramachandran, V.S. (2009). Reflections on the Mirror Neuron System: Their Evolutionary Functions Beyond Motor Representation . In Pineda, J.A. Mirror Neuron Systems: The Role of Mirroring Processes in Social Cognition. Humana Press. pp. 39–62. ISBN 978-1-934115-34-3.

119. Ramachandran, V.S. (1 January 2009). Self Awareness: The Last Frontier, Edge Foundation web essay . Retrieved 26 July 2011.

120. Nichols, S.; Grantham, T. (2000). Adaptive Complexity and Phenomenal Consciousness . Philosophy of Science. 67 *(4): 648–670.* doi:10.1086/392859. JSTOR 188711.

121. Freeman and Herron. Evolutionary Analysis. 2007. Pearson Education, NJ.

122. Peter *Århem*; B. I. B. Lindahl; Paul R. Manger & Ann B. Butler (2008). On the origin of consciousness—some amniote scenarios . In Hans Liljenström & Peter *Århem*. Consciousness Transitions: Phylogenetic, Ontogenetic, and Physiological Aspects. Elsevier. ISBN 978-0-444-52977-0.

123. Eccles, J. C. (1992). Evolution of consciousness . Proceedings of the National Academy of Sciences of the United States of America. 89 *(16): 7320–7324. doi:10.1073/pnas.89.16.7320. JSTOR 2360081. PMC 49701. PMID 1502142.*

124. Peters, Frederic Consciousness as Recursive, Spatiotemporal Self-Location

125. Baars, Bernard J. A Cognitive Theory of Consciousness. 1993. Cambridge University Press.

126. Gaulin, Steven J. C. and Donald H. McBurney (2003) Evolutionary psychology. Prentice Hall, pp. 101–121. ISBN 978-0-13-111529-3

127. Edelman, G. (2004). Wider than the sky: The phenomenal gift of consciousness: Yale Univ Pr.

128. Edelman, G. M. (1992). Bright Air, Brilliant Fire: On the Matter of the Mind. Basic Books, New York.

129. c Merker, B (2007). Consciousness without a cerebral cortex: A challenge for neuroscience and medicine . Behavioral and Brain Sciences. 30 *(01): 63–81.* doi:10.1017/s0140525x07000891.

130. Edelman, G. M. (2004). Wider than the sky: a revolutionary view of consciousness. Penguin Press Science, London, UK.

131. Edelman, G. M. (200). Second Nature:Brain Science and Human Knowledge. Yale University Press, New Haven, CT.

132. Hobson, J. A. (2009). REM sleep and dreaming: towards a theory of protoconsciousness (PDF). Nature Reviews Neuroscience. 10 *(11): 803–862.* doi:10.1038/nrn2716. PMID 19794431.

133. Voss, Ursula (2010). Changes in EEG pre and post awakening . International Review of Neurobiology. 93: 23–56. doi:10.1016/S0074-7742(10)93002-X.

134. Kuiken, Don (2010). Primary and secondary consciousness during dreaming . International Journal of Dream Research. 3 *(1): 21–25.*

135. Low, Philip et al. (2012) The Cambridge Declaration on ConsciousnessPublicly proclaimed in Cambridge, UK, on July 7, 2012, at the Francis Crick Memorial Conference on Consciousness in Human and non-Human Animals.

136. "Consciousness in Human and non-Human Animals" Francis Crick Memorial Conference, 7 July 2012, Cambridge, UK.

137. Campbell, C.B.G.; Hodos, W. (1991). The Scala Naturae revisited: Evolutionary scales and anagenesis in comparative psychology . J. Comp. Psychol. 105 *(3): 211–221. doi:10.1037/0735-7036.105.3.211.*

138. Shettleworth (1998).

139. Reznikova (2007).

140. Dogs (and probably many other animals) have a conscience too! . en.tsu.ru. Retrieved 2016-12-05.

141. Nkisi Audio Text 1

142. Maria Armental (8 May 2008). Jane Goodall attends opening of bird sanctuary . www.projo.com. Archived from the original on 29 June 2011.

143. Gill, Victoria. Parrots choose to work together . BBC Nature News. Archivedfrom the original on 20 May 2011. Retrieved 19 May 2011.

144. What is this octopus thinking?. By Garry Hamilton.

145. NFW.org?, Is the octopus really the invertebrate intellect of the sea, by Doug Stewart. In: National Wildlife. Feb/Mar 1997, vol.35 no.2.

146. Giant Octopus—Mighty but Secretive Denizen of the Deep

147. Slate.com, How Smart is the Octopus?

148. Yekutieli, Yoram; Sagiv-Zohar, Roni; Aharonov, Ranit; Enge, Yaakov; Hochner, Binyamin; Flash, Tamar (2005). Dynamic Model of the Octopus Arm. I. Biomechanics of the Octopus Reaching Movement . J. Neurophysiology. 94 *(2): 1443–1458. doi:10.1152/jn.00684.2004. PMID 15829594.*

149. Zullo, L; Sumbre, G; Agnisola, C; Flash, T; Hochner, B (2009). Nonsomatotopic organization of the higher motor centers in octopus . Curr Biol. 19 *(19): 1632–6. doi:10.1016/j.cub.2009.07.067. PMID 19765993.*

150. Octopus intelligence: jar opening

151. What behavior can we expect of octopuses?. By Dr. Jennifer Mather, Department of Psychology and Neuroscience, University of Lethbridge and Roland C. Anderson, The Seattle Aquarium.

152. Octopus snatches coconut and runs . BBC News. 14 December 2009. Retrieved 20 May 2010.

153. http://www.edutube.org/video/coconut-shelter-evidence-tool-use-octopuses

CHAPTER 3

Animal Language

Chapter Contents

All animals use some form of communication, although some animal communication is more complex than others. Animal language is any form of communication that shows similarities to human language; however, there are significant differences. Some animals use signs, signals, or sounds to communicate. Lexigrams, or figures and symbols that represent words, are commonly used by chimpanzees and baboons, while animals such as birds and whales use song to communicate among one another. Bees uses complex dances to convey information about location. Other animals use odors or body movements to communicate.

INTRODUCTION

Animal languages are forms of non-human animal communication that show similarities to human language. Animals communicate by using a variety of signs such as sounds or movements. Such signing may be considered complex enough to be called a form of language if the inventory of signs is large, the signs are relatively arbitrary, and the animals seem to produce them with a degree of volition (as opposed to relatively automatic conditioned behaviors or unconditioned instincts, usually including facial expressions). In experimental tests, animal communication may also be evidenced through the use of lexigrams (as used by chimpanzees and bonobos). While the term animal language is widely used, researchers agree that animal languages are not as complex or expressive as human language.

Many researchers argue that animal communication lacks a key aspect of human language, that is, the creation of new patterns of signs under varied circumstances. (In contrast, for example, humans routinely produce entirely new combinations of words.) Some researchers, including the linguist Charles Hockett, argue that human language and animal communication differ so much that the underlying principles are unrelated. Accordingly, linguist Thomas A. Sebeok has proposed to not use the term "language" for animal sign systems. Marc Hauser, Noam Chomsky, and W. Tecumseh Fitch assert an evolutionary continuum exists between the communication methods of animal and human language.

Human and chimp, in this case Claudine André with a bonobo.

The following properties of human language have been argued to separate it from animal communication:

- **Arbitrariness**: there is usually no rational relationship between a sound or sign and its meaning. For example, there is nothing intrinsically "housy" about the word "house".
- **Discreteness:** language is composed of small, repeatable parts (discrete units) that are used in combination to create meaning.
- **Displacement:** languages can be used to communicate ideas about things that are not in the immediate vicinity either spatially or temporally.
- **Duality of patterning:** the smallest meaningful units (words, morphemes) consist of sequences of units without meaning. This is also referred to as **double articulation**.
- **Productivity:** users can understand and create an indefinitely large number of utterances.
- **Semanticity**: specific signals have specific meanings.

Research with apes, like that of Francine Patterson with Koko (gorilla) or Allen and Beatrix Gardner with Washoe (chimpanzee), suggested that apes are capable of using language that meets some of these requirements such as arbitrariness, discreteness, and productivity.

In the wild, chimpanzees have been seen "talking" to each other when warning about approaching danger. For example, if one chimpanzee sees a snake, he makes a low, rumbling noise, signaling for all the other chimps to climb into nearby trees. In this case, the chimpanzees' communication does not indicate displacement, as it is entirely contained to an observable event.

Arbitrariness has been noted in meerkat calls; bee dances demonstrate elements of spatial displacement; and cultural transmission has possibly occurred between the celebrated bonobos Kanzi and Panbanisha.

Human language may not be completely arbitrary. Research has shown that almost all humans naturally demonstrate limited crossmodal perception (e.g. synesthesia) and multisensory integration, as illustrated by the Kiki and Booba study. Other recent research has tried to explain how the structure of human language emerged, comparing two different aspects of hierarchical structure present in animal communication and proposing that human language arose out of these two separate systems.

Claims that animals have language skills akin to humans however, are extremely controversial. As Steven Pinker illustrates in his book *The Language Instinct*, claims that chimpanzees can acquire language are exaggerated and rest on very limited or specious data.

The American linguist Charles Hockett theorized that there are sixteen features of human language that distinguished human communication from that of animals. He called these the design features of language. The features mentioned below have so far been found in all spoken human languages and at least one is missing from all other animal communication systems.

- Vocal-auditory channel: sounds emitted from the mouth and perceived by the auditory system. This applies to many animal communication systems, but there are many exceptions. Ex. An alternative to vocal-auditory communication is visual communication. An example is cobras extending the ribs behind their heads to send the message of intimidation or of feeling threatened. In humans, sign languages provide many examples of fully formed languages that use a visual channel.
- Broadcast transmission and directional reception: this requires that the recipient can tell the direction that the signal comes from and thus the originator of the signal.
- Rapid fading (transitory nature): Signal lasts a short time. This is true of all systems involving sound. It does not take into account audio recording technology and is also not true for written language. It tends not to apply to animal signals involving chemicals and smells which often fade slowly. For example, a skunk's smell, produced in its glands, lingers to deter a predator from attacking.
- Interchangeability: All utterances that are understood can be produced. This is different from some communication systems where, for example, males produce one set of behaviours and females another and they are unable to interchange these messages so that males use the female signal and vice versa. For example, Heliothine moths have differentiated communication: females are able to send a chemical to indicate preparedness to mate, while males cannot send the chemical.

- Total feedback: The sender of a message is aware of the message being sent.
- Specialization: The signal produced is intended for communication and is not due to another behavior. For example, dog panting is a natural reaction to being overheated, but is not produced to specifically relay a particular message.
- Semanticity: There is some fixed relationship between a signal and a meaning.

PRIMATE: STUDIED EXAMPLES

Humans are able to distinguish real words from fake words based on the phonological order of the word itself. In a 2013 study, baboons have been shown to have this skill, as well. The discovery has led researchers to believe that reading is not as advanced a skill as previously believed, but instead based on the ability to recognize and distinguish letters from one another. The experimental setup consisted of six young adult baboons, and results were measured by allowing the animals to use a touch screen and selecting whether or not the displayed word was indeed a real word, or a nonword such as dran or telk. The study lasted for six weeks, with approximately 50,000 tests completed in that time. The experimenters explain the use of bigrams, which are combinations of two (usually different) letters. They tell us that the bigrams used in nonwords are rare, while the bigrams used in real words are more common. Further studies will attempt to teach baboons how to use an artificial alphabet.

In a 2016 study, a team of biologists from several universities concluded that macaques possess vocal tracts physically capable of speech, "but lack a speech-ready brain to control it".

NON-PRIMATES: STUDIED EXAMPLES

Among the most studied examples of animal languages are:

Birds

- Bird songs: Songbirds can be very articulate. African grey parrots are famous for their ability to mimic human language, and at least one specimen, Alex, appeared able to answer a number of simple questions about objects he was presented with. Parrots, hummingbirds and songbirds - display vocal learning patterns.

Insects

- Bee dance: Used to communicate direction and distance of food source in many species of bees.

Mammals

- African forest elephants: Cornell University's Elephant Listening Project began in 1999 when Katy Payne began studying the calls of African forest elephants in Dzanga National Park in the Central African Republic. Andrea Turkalo has continued Payne's work in Dzanga National Park observing elephant communication. For nearly 20 years, Turkalo has spent the majority of her time using a spectrogram to record the noises that the elephants make. After extensive observation and research, she has been able to recognize elephants by their voices. Researchers hope to translate these voices into an elephant dictionary, but that will likely not occur for many years. Because elephant calls are often made at very low frequencies, this spectrogram is able to detect lower frequencies that human ears are unable to hear, allowing Turkalo to get a better idea of what she perceives the elephants to be saying. Cornell's research on African forest elephants has challenged the idea that humans are considerably better at using language and that animals only have a small repertoire of information that they can convey to others. As Turkalo explained on *60 Minutes*' "The Secret Language of Elephants," "Many of their calls are in some ways similar to human speech."

- Mustached bats: Since these animals spend most of their lives in the dark, they rely heavily on their auditory system to communicate. This acoustic communication includes echolocation or using calls to locate each other in the darkness. Studies have shown that mustached bats use a wide variety of calls to communicate with one another. These calls include 33 different sounds, or "syllables," that the bats then either use alone or combine in various ways to form "composite" syllables.
- Prairie dogs: Dr. Con Slobodchikoff studied prairie dog communication and discovered:
    - different alarm calls for different species of predators;
    - different escape behaviors for different species of predators;
    - transmission of semantic information, in that playbacks of alarm calls in the absence of predators lead to escape behaviors that are appropriate to the type of predator which elicited the alarm calls;
    - alarm calls containing descriptive information about the general size, color, and speed of travel of the predator.

Aquatic Mammals

- Bottlenose dolphins: Dolphins can hear one another up to 6 miles apart underwater. In one National Geographic article, the success of a mother dolphin communicating with her baby using a telephone was outlined. Researchers noted that it appeared that both dolphins knew who they were speaking with and what they were speaking about. Not only do dolphins communicate via nonverbal cues, but they also seem to chatter and respond to other dolphin's vocalizations.

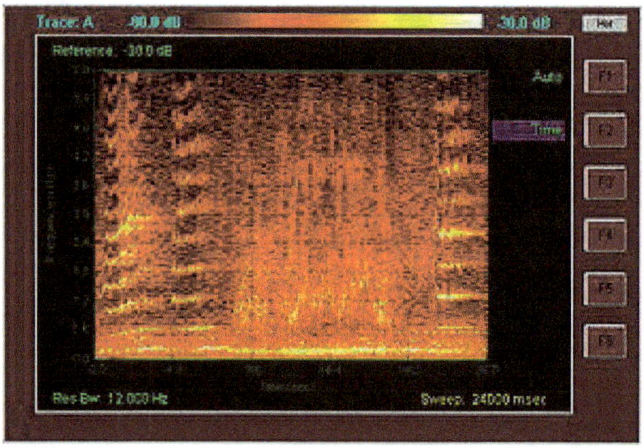

Spectrogram of humpback whale vocalizations. Detail is shown for the first 24 seconds of the 37 second recording humpback whale "song". The ethereal whale "songs" and echolocation "clicks" are visible as horizontal striations and vertical sweeps respectively.

- Whales: Two groups of whales, the humpback whale and a subspecies of blue whale found in the Indian Ocean, are known to produce repetitious sounds at varying frequencies known as whale song. Male humpback whales perform these vocalizations only during the mating season, and so it is surmised the purpose of songs is to aid sexual selection. Humpbacks also make a sound called a feeding call, five to ten seconds in length of near constant frequency. Humpbacks generally feed cooperatively by gathering in groups, swimming underneath shoals of fish and all lunging up vertically through the fish and out of the water together. Prior to these lunges, whales make their feeding call. The exact purpose of the call is not known, but research suggests that fish react to it. When the sound was played back to them, a group of herring responded to the sound by moving away from the call, even though no whale was present.
- Sea lions: Beginning in 1971 and continuing until present day, Dr. Ronald J. Schusterman and his research associates have studied sea lions' cognitive ability. They have discovered that sea lions are able to recognize relationships between stimuli based on similar functions or connections made with their peers, rather than only the stimuli's common features. This is called "equivalence classification." This ability to recognize equivalence may be a precursor to language. Research is currently being conducted at the Pinniped Cognition & Sensory Systems Laboratory to determine how sea lions form these equivalence relationships. Sea lions have also been proven to be able to understand simple syntax and commands when taught an artificial sign language similar to the one used with primates. The sea lions studied were able to learn and use a number of syntactic relations between the signs they were taught, such as how the signs should be arranged in relation to each other. However, the sea lions rarely used the signs semantically or logically. In the wild it's thought that sea lions use the reasoning skills associated with equivalence classification in

order to make important decisions that can affect their rate of survival (e.g. recognizing friends and family or avoiding enemies and predators). Sea lions use the following to display their language:

- Sea lions use their bodies in various postural positions to display communication.
- Sea lion's vocal cords limit their ability to convey sounds to a range of barks, chirps, clicks, moans, growls and squeaks.
- There has yet to be an experiment which proves for certain that sea lions use echolocation as a means of communication.

The effects of learning on auditory signaling in these animals is of special interest. Several investigators have pointed out that some marine mammals appear to have an extraordinary capacity to alter both the contextual and structural features of their vocalizations as a result of experience. Janik and Slater (2000) have stated that learning can modify the emission of vocalizations in one of two ways: (1) by influencing the context in which a particular signal is used and/or (2) by altering the acoustic structure of the call itself. Male California sea lions can learn to inhibit their barking in the presence of any male dominant to them, but vocalize normally when dominant males are absent. Recent work on gray seals show different call types can be selectively conditioned and placed under biased control of different cues (Schusterman, in press) and the use of food reinforcement can also modify vocal emissions. "Hoover", a captive male harbor seal demonstrated a convincing case of vocal mimicry. However similar observations have not been reported since. Still shows under the right circumstances pinnipeds may use auditory experience, in addition to environmental consequences such as food reinforcement and social feedback to modify their vocal emissions.

In a 1992 study, Robert Gisiner and Ronald J. Schusterman conducted experiments in which they attempted to teach Rocky, a female California sea lion, syntax.Rocky was taught signed words, then she was asked to perform various tasks dependent on word order after viewing a signed instruction.It was found that Rocky was able to determine relations between signs and words, and form a basic form of syntax. A 1993 study by Ronald J Schusterman and David Kastak found that the California sea lion was capable of understanding abstract concepts such as symmetry, sameness and transitivity. This provides a strong backing to the theory that Equivalence relations can form without language.

The distinctive sound of sea lions is produced both above and below water. To mark territory, sea lions "bark", with non-alpha males making more noise than alphas. Although females also bark, they do so less frequently and most often in connection with birthing pups or caring for their young. Females produce a highly directional bawling vocalization, the pup attraction call, which helps mother and pup locate one another. As noted in *Animal Behavior*, their amphibious lifestyle has made them need acoustic communication for social organization while on land.

Sea lions can hear frequencies as low as 100 Hz and as high as 40,000 Hz and vocalize between the ranges of 100 to 10,000 Hz.

Mollusks

- Caribbean reef squid have been shown to communicate using a variety of color, shape, and texture changes. Squid are capable of rapid changes in skin color and pattern through nervous control of chromatophores. In addition to camouflage and appearing larger in the face of a threat, squids use color, patterns, and flashing to communicate with one another in various courtship rituals. Caribbean reef squid can send one message via color patterns to a squid on their right, while they send another message to a squid on their left.

COMPARISON OF THE TERMS "ANIMAL LANGUAGE" AND "ANIMAL COMMUNICATION"

It is worth distinguishing "animal language" from "animal communication", although there is some comparative interchange in certain cases (e.g. Cheney & Seyfarth's vervet monkey call studies). Thus "animal language" typically does not include bee dancing, bird song, whale song, dolphin signature whistles, prairie dogs, nor the communicative systems found in most social mammals. The features of language as listed above are a dated formulation by Hockett in 1960. Through this formulation Hockett made one of the earliest attempts to break down features of human language for the purpose of applying Darwinian gradualism. Although an influence on early animal language efforts (see below), is today not considered the key architecture at the core of animal language research.

Animal Language results are controversial for several reasons. (For a related controversy, see also Clever Hans.) In the 1970s John Lilly was attempting to "break the code": to fully communicate ideas and concepts with wild populations of dolphins so that we could "speak" to them, and share our cultures, histories, and more. This effort failed. Early chimpanzee work was with chimpanzee infants raised as if they were human; a test of the nature vs. nurture hypothesis. Chimpanzees have a laryngeal structure very different from that of humans, and it has been suggested that chimpanzees are not capable of voluntary control of their breathing, although better studies are needed to accurately confirm this. This combination is thought to make it very difficult for the chimpanzees to reproduce the vocal intonations required for human language. Researchers eventually moved towards a gestural (sign language) modality, as well as keyboard devices laden with buttons adorned with symbols (known as lexigrams) that the animals could press to produce artificial language. Other chimpanzees learned by observing human subjects performing the task. This latter group of researchers studying chimpanzee communication through symbol recognition (keyboard) as well as through the use of sign language (gestural), are on the forefront of communicative breakthroughs in the study of animal language, and they are familiar with their subjects on a first name basis: Sarah, Lana, Kanzi, Koko, Sherman, Austin and Chantek.

Perhaps the best known critic of Animal Language is Herbert Terrace. Terrace's 1979 criticism using his own research with the chimpanzee Nim Chimpskywas scathing and basically spelled the end of animal language research in that era, most of which emphasized the production of language by animals. In short, he accused researchers of over-interpreting their results, especially as it is rarely parsimonious to ascribe true intentional "language production" when other simpler explanations for the behaviors (gestural hand signs) could be put forth. Also, his animals failed to show generalization of the concept of reference between the modalities of comprehension and production; this generalization is one of many fundamental ones that are trivial for human language use. The simpler explanation according to Terrace was that the animals had learned a sophisticated series of context-based behavioral strategies to obtain either primary (food) or social reinforcement, behaviors that could be over-interpreted as language use.

In 1984 during this anti-Animal Language backlash, Louis Herman published an account of artificial language in the bottlenosed dolphin in the journal Cognition.A major difference between Herman's work and previous research was his emphasis on a method of studying language comprehension only (rather than language comprehension and production by the animal(s)), which enabled rigorous controls and statistical tests, largely because he was limiting his researchers to evaluating the animals' physical behaviors (in response to sentences) with blinded observers, rather than attempting to interpret possible language utterances or productions. The dolphins' names here were Akeakamai and Phoenix. Irene Pepperberg used the vocal modality for language production and comprehension in an African grey parrot named Alex in the verbal mode, and Sue Savage-Rumbaugh continues to study bonobos such as Kanzi and Panbanisha. R. Schusterman duplicated many of the dolphin results in his California sea lions (Rocky), and came from a more behaviorist tradition than Herman's cognitive approach. Schusterman's emphasis is on the importance on a learning structure known as equivalence classes.

However, overall, there has not been any meaningful dialog between the linguistics and animal language spheres, despite capturing the public's imagination in the popular press. Also, the growing field of language evolution is another source of future interchange between these disciplines. Most primate researchers tend to show a bias toward a shared pre-linguistic ability between humans and chimpanzees, dating back to a common ancestor, while dolphin and parrot researchers stress the general cognitive principles underlying these abilities. More recent related controversies regarding animal abilities include the closely linked areas of Theory of mind, Imitation (e.g. Nehaniv & Dautenhahn, 2002), Animal Culture (e.g. Rendell & Whitehead, 2001), and Language Evolution (e.g. Christiansen & Kirby, 2003).

There has been a recent emergence in animal language research which has contested the idea that animal communication is less sophisticated than human communication. Denise Herzing has done research on dolphins in the Bahamas whereby she created a two-way conversation via a submerged keyboard. The keyboard allows divers to communicate with wild dolphins. By using sounds and symbols on each key the dolphins could either press the key with their nose or mimic the whistling sound emitted in order to ask humans for a specific prop. This ongoing experiment has shown that in non-linguistic creatures brilliant and rapid thinking does occur despite our previous conceptions of animal communication. Further research done with Kanzi using lexigrams has strengthened the idea that animal communication is much more complex then we once thought.

REFERENCES

1. Hockett, Charles F. (1960). "Logical considerations in the study of animal

communication". In Lanyon, W.E.; Tavolga, W.N. Animals sounds and animal communication. American Institute of Biological Sciences. pp. 392–430.

2. Martinelli, Dario (2010). "Introduction to Zoosemiotics". Biosemiotics. 5: 1–64. doi:10.1007/978-90-481-9249-6_1.
3. Hauser, Marc D.; Chomsky, Noam; Fitch, W. Tecumseh (22 November 2002). "The Faculty of Language: What Is It, Who Has It, and How Did It Evolve?" (PDF). Science. American Association for the Advancement of Science. pp. 1569–1579. Retrieved 28 March 2014. We argue that an understanding of the faculty of language requires substantial interdisciplinary cooperation. We suggest how current developments in linguistics can be profitably wedded to work in evolutionary biology, anthropology, psychology, and neuroscience. We submit that a distinction should be made between the faculty of language in the broad sense (FLB)and in the narrow sense (FLN). FLB includes a sensory-motor system, a conceptual-intentional system, and the computational mechanisms for recursion, providing the capacity to generate an infinite range of expressions from a finite set of elements. We hypothesize that FLN only includes recursion and is the only uniquely human component of the faculty of language. We further argue that FLN may have evolved for reasons other than language, hence comparative studies might look for evidence of such computations outside of the domain of communication (for example, number, navigation, and social relations).
4. Denham, Kristin; Lobeck, Anne (2010). Linguistics for Everyone: An Introduction (Instructor's ed.). Wadsworth, Cengage Learning. pp. 4–5. ISBN 9781428205833.
5. Fitch, WT. (Feb 2011). "Unity and diversity in human language.". Philos Trans R Soc Lond B Biol Sci. 366 (1563): 376–88. doi:10.1098/rstb.2010.0223. PMC 3013471Freely accessible. PMID 21199842.
6. Patterson, Francine.; Linden, Eugene. (1981). The education of Kok. New York: Holt. ISBN 978-0-03-046101-9. OCLC 7283799.
7. Gardner, R. A.; Gardner, B. T. (1969). "Teaching Sign Language to a Chimpanzee". Science. 165 (3894): 664–672. doi:10.1126/science.165.3894.664. ISSN 0036-8075. PMID 5793972.
8. Gardner, B.T.; Gardner, R.A. (1975). "Evidence for sentence constituents in the early utterances of child and chimpanzee". Journal of Experimental Psychology General. 104 (3): 244–267. doi:10.1037/0096-3445.104.3.244.
9. Fernández, Eva M.; Cairns, Helen Smith. (2011). Fundamentals of psycholinguistic. Chichester, West Sussex [England] ; Malden, MA: Wiley-Blackwell. ISBN 978-1-4051-9147-0. OCLC 667883441.
10. Raffaele, P (November 2006). "Speaking Bonobo". Simithsonian. Retrieved 2008-03-18.
11. Maurer, D.; Pathman, T.; Mondloch, CJ. (May 2006). "The shape of boubas: sound-shape correspondences in toddlers and adults.". Dev Sci. 9 (3): 316–22. doi:10.1111/j.1467-7687.2006.00495.x. PMID 16669803.
12. Kubovy, M.; Yu, M. (Apr 2012). "Multistability, cross-modal binding and the additivity of conjoined grouping principles.". Philos Trans R Soc Lond B Biol

Sci. 367 (1591): 954–64. doi:10.1098/rstb.2011.0365. PMC 3282311Freely accessible. PMID 22371617.

13. Miyagawa, Shigeru. "The emergence of hierarchical structure in human language". Frontiers in Language Sciences. Retrieved 2013-02-24.
14. Pinker, Steven (2000). The language instinct : how the mind creates language. New York: Perennial Classics. ISBN 9780060958336. OCLC 45992871.
15. Cobra Master (2003-12-12). "Introduction". Cobras.org. Retrieved 2013-05-19.
16. "Skunks, Skunk Pictures, Skunk Facts - National Geographic". Animals. nationalgeographic.com. 2013-05-15. Retrieved 2013-05-19.
17. "Chemical communication in heliothine moths". Journal of Comparative Physiology A. 169. 1991-09-01. doi:10.1007/BF00206989. Retrieved 2013-05-19.
18. Haghighat, Leila. "Baboons Can Learn to Recognize Words". Nature News. Retrieved 15 April 2013.
19. Fitch, W. T.; de Boer, B.; Mathur, N.; Ghazanfar, A. A. (December 2016). "Monkey vocal tracts are speech-ready". Science Advances. 2 (12). doi:10.1126/sciadv.1600723.
20. "Why can't monkeys talk? Their anatomy is 'speech-ready' but their brains aren't wired for it: neuroscientist". National Post. Retrieved 10 December 2016.
21. "The Elephant Listening Project". Cornell University. Retrieved 28 December 2013.
22. "The Elephant Listening Project". Cornell University Lab. Retrieved 2013-02-24.
23. "The Secret Language of Elephants". CBS News 60 Minutes. Retrieved 2013-02-24.
24. Kanwal, J.S.; Matsumura, S.; Ohlemiller, K.; Suga, N. (1994). "Analysis of acoustic elements and syntax in communication sounds emitted by mustached bats". Journal of the Acoustic Society of America. 94 (3): 1229–1254.
25. "Con Slobodchikoff". nau.edu. Retrieved 10 December 2016.
26. "Dolphins' Secret Language". Young Peoples Trust for the Environment. 29 February 2012. Archived from the original on 27 August 2012.
27. "The Secret Language of Dolphins". National Geographic Kids. Retrieved 2013-03-31.
28. "NOVA scienceNOW: Smart Marine Mammals - Smart Sea Lions". teachersdomain.org. Retrieved 10 December 2016.
29. http://www.pinnipedlab.org/ The Pinniped Cognition & Sensory Systems Laboratory
30. Gisiner, Robert; Schusterman, Ronald J. (1992). "Sequence, syntax, and semantics: Responses of a language-trained sea lion (Zalophus californianus) to novel sign combinations." (PDF). Journal of Comparative Psychology. 106 (1): 78–91. doi:10.1037/0735-7036.106.1.78.
31. "CALIFORNIA SEA LION - Communication". Seaworld.org. Retrieved 2013-05-19.

32. "Sea Lion Info - Dolphin Research Center". dolphins.org. Retrieved 10 December 2016.
33. Janik, VM.; Slater, PJ. (Jul 2000). "The different roles of social learning in vocal communication.". Anim Behav. 60 (1): 1–11. doi:10.1006/anbe.2000.1410. PMID 10924198.
34. Schusterman, RJ.; Dawson, RG. (Apr 1968). "Barking, dominance, and territoriality in male sea lions.". Science. 160 (3826): 434–6. doi:10.1126/science.160.3826.434. PMID 5689412.
35. Shapiro, AD.; Slater, PJ.; Janik, VM. (Dec 2004). "Call usage learning in gray seals (Halichoerus grypus).". J Comp Psychol. 118 (4): 447–54. doi:10.1037/0735-7036.118.4.447. PMID 15584781.
36. "Oceanography". Onr.navy.mil. Retrieved 2013-05-19.
37. Cloney, RA; Florey, E (1968). "Ultrastructure of cephalopod chromatophore organs". Zeitschrift für Zellforschung und mikroskopische Anatomie. 89 (2): 250–80. doi:10.1007/BF00347297. PMID 5700268.
38. "Sepioteuthis sepioidea; Caribbean Reef squid". The Cephalopod Page. Retrieved 2013-05-19.
39. Byrne, R.A., U. Griebel, J.B. Wood & J.A. Mather 2003. "Squids say it with skin: a graphic model for skin displays in Caribbean Reef Squid." (PDF). (3.86 MB) Berliner Geowissenschaftliche Abhandlungen 3: 29-35.
40. Seyfarth, R. M.; Cheney, D.L. (1990). "The assessment by vervet monkeys of their own and other species' alarm calls". Animal Behaviour. 40 (4): 754–764. doi:10.1016/S0003-3472(05)80704-3.
41. Terrace, Herbert S. (1979). Nim. New York: Knopf : distributed by Random House. ISBN 978-0-394-40250-5. OCLC 5102119.
42. Terrace, H.S.; Petitto, L.A.; Sanders, R.J.; Bever, T.G. (1979). "Can an ape create a sentence?". Science. 206 (4421): 891–902. doi:10.1126/science.504995. PMID 504995.
43. Herman, L. M.; Richards, D. G.; Wolz, J. P. (1984). "Comprehension of sentences by bottlenosed dolphins". Cognition. 16 (2): 129–219. doi:10.1016/0010-0277(84)90003-9. PMID 6540652.
44. Pepperberg, Irene M. (1999). The Alex studies : cognitive and communicative abilities of grey parrot. Cambridge, Mass.: Harvard University Press. ISBN 978-0-674-00051-3. OCLC 807730081.
45. Pepperberg, IM. (Oct 2010). "Vocal learning in Grey parrots: A brief review of perception, production, and cross-species comparisons.". Brain Lang. 115 (1): 81–91. doi:10.1016/j.bandl.2009.11.002. PMID 20199805.
46. Pepperberg, IM.; Carey, S. (Nov 2012). "Grey parrot number acquisition: the inference of cardinal value from ordinal position on the numeral list.". Cognition. 125 (2): 219–32. doi:10.1016/j.cognition.2012.07.003. PMC 3434310Freely accessible. PMID 22878117.

47. Pepperberg, IM. (Feb 2013). "Abstract concepts: data from a Grey parrot.". Behav Processes. 93: 82–90. doi:10.1016/j.beproc.2012.09.016. PMID 23089384.

48. Savage-Rumbaugh, E. S. (1990). "Language Acquisition in a Nonhuman Species: Implications for the innateness debate". Developmental Psychobiology. 23 (7): 599–620. doi:10.1002/dev.420230706.

49. Savage-Rumbaugh, E. S.; Fields, W. M. (2000). "Linguistic, cultural and cognitive capacities of bonobos (Pan paniscus)". Culture and Psychology. 6 (2): 131–154. doi:10.1177/1354067X0062003.

50. Schusterman, RJ.; Kastak, D. (May 1998). "Functional equivalence in a California sea lion: relevance to animal social and communicative interactions.". Anim Behav. 55 (5): 1087–95. doi:10.1006/anbe.1997.0654. PMID 9632496.

51. Kastak, CR.; Schusterman, RJ.; Kastak, D. (Sep 2001). "Equivalence classification by California sea lions using class-specific reinforcers.". J Exp Anal Behav. 76 (2): 131–58. doi:10.1901/jeab.2001.76-131. PMC 1284831Freely accessible. PMID 11599636.

52. Nehaniv, Chrystopher; Dautenhahn, Kerstin (2002). Imitation in animals and artifacts. Cambridge, Mass: MIT Press. ISBN 9780262271219. OCLC 51938434.

53. Rendell, L.; Whitehead, H. (2001). "Culture in whales and dolphins". Beharioral and Brain Sciences. 24 (2): 309–382. doi:10.1017/S0140525X0100396X. PMID 11530544.

54. Christiansen, Morten H.; Kirby, Simon. (2003). Language evolution. Oxford ; New York: Oxford University Press. ISBN 978-0-19-924484-3. OCLC 51235137.

55. Savage-Rumbaugh, S.; Rumbaugh, D.; Fields, W. "Empirical kanzi: The ape language controversy revisited. (2009)". Skeptic. 15 (1): 25–33.

CHAPTER 4

Elephant Cognition

Chapter Contents

INTRODUCTION

Most contemporary ethologists view elephants as one of the world's most intelligent animals. With a mass of just over 5 kg (11 lb), elephant brains have more mass than those of any other land animal, and although the largest whales have body masses twenty-fold those of a typical elephant, whale brains are barely twice the mass of an elephant's brain. In addition, elephants have a total of 300 billion neurons. Elephant brains are similar to humans' in terms of general connectivity and areas. The elephant cortex has as many neurons as a human brain, suggesting convergent evolution.

Elephants manifest a wide variety of behaviors, including those associated with grief, learning, mimicry, play, altruism, use of tools, compassion, cooperation,self-awareness, memory, and communication. Further, evidence suggests elephants may understand pointing: the ability to nonverbally communicate an object by extending a finger, or equivalent. Such behaviors suggest that elephants are highly intelligent; it is thought they are equal with cetaceans and primates in this regard. Due to such claims of high intelligence and due to strong family ties of elephants, some researchers argue it is morally wrong for humans to cull them. The Ancient Greek philosopher, Aristotle, once said that the elephant was the animal which surpasses all others in wit and mind.

However, a few elephant researchers, and some ethologists, point to experimental and anecdotal evidence which appear to contradict the view that elephants are self-aware, can think, and possess a theory of mind.

BRAIN STRUCTURE

Cerebral Cortex

The elephant (both Asian and African) has a very large and highly complex neocortex, a trait also shared by humans, apes and certain dolphin species.

Asian elephants have the greatest volume of cerebral cortex available for cognitive processing of all existing land animals. Elephants have a volume of cerebral cortex available for cognitive processing that exceeds that of any primate species, with one study suggesting elephants be placed in the category of great apes in terms of cognitive abilities for tool use and tool making.

The elephant brain exhibits a gyral pattern more complex and with more numerous convolutions, or brain folds, than that of humans, primates, or carnivores, but less complex than cetaceans, although elephants have as many cortical neurons (nerve cells) and cortical synapses as that of humans, which is more than that of the cetaceans. Elephants are believed to rank equal with dolphins in terms of problem-solving abilities, and many scientists tend to rank elephant intelligence at the same level as cetaceans; a 2011 article published by ABC Science states that, "elephants [are as] smart as chimps, [and] dolphins.

Other Features of the Brain

Elephants also have a very large and highly convoluted hippocampus, a brain structure in the limbic system that is much bigger than that of any human, primate or cetacean. The hippocampus of an elephant takes up about 0.7% of the central structures of the brain, comparable to 0.5% for humans and with 0.1% in Risso's dolphins and 0.05% in bottlenose dolphins. The hippocampus is linked to emotion through the processing of certain types of memory, especially spatial. This is thought to be possibly why elephants suffer from psychological flashbacks and the equivalent of post-traumatic stress disorder (PTSD).

The encephalization quotient (EQ) (the size of the brain relative to body size) of elephants ranges from 1.13 to 2.36. The average EQ is 2.14 for Asian elephants, and 1.67 for African, with the overall average being 1.88.[:151] In comparison to other animals, the La Plata dolphin has an EQ of 1.67; the Ganges River dolphin of 1.55; the orca of 2.57; the bottlenose dolphin of 4.14; and the tucuxi dolphin of 4.56; chimpanzees at 2.49; dogs at 1.17, cats at 1.00; and mice at 0.50. Humans have an EQ of 7.44.

Brain size at Birth Relative to Adult Brain Size

Like humans, elephants must learn behavior as they grow up. They are not born with the instincts of how to survive. Elephants have a very long period in their lives for learning, lasting for around ten years. One comparative way to try to gauge intelligence is to compare brain size at birth to the fully developed adult brain. This indicates how much learning a species accumulates while young. The majority of mammals are born with a brain close to 90% of the adult weight, while humans are born with 28%, bottlenose dolphins with 42.5%, chimpanzees with 54%, and elephants with 35%. This indicates that elephants have the highest amount of learning to undergo next to humans, and behavior is not mere instinct but must be taught throughout life. It should be noted that instinct is quite different from learned intelligence. Parents teach their young how to feed, use tools and learn their place in the highly complex elephant society. The cerebrumtemporal lobes, which function as storage of memory, are much larger than those of a human.

Spindle Neurons

Spindle cells appear to play a central role in the development of intelligent behavior. As well as in humans and the great apes, spindle neurons are also found in the brains of both Asian and African elephants, as well as humpback whales, fin whales, killer whales, sperm whales, bottlenose dolphins, Risso's dolphins, and beluga whales. The remarkable similarity between the elephant brain and the human brain supports the thesis of convergent evolution.

ELEPHANT SOCIETY

The elephant has one of the most closely knit societies of any living species. Elephant families can only be separated by death or capture. Cynthia Moss, an ethologist specialising in elephants, recalls an event involving a family of African elephants. Two members of the family were shot by poachers, who were subsequently chased off by the remaining elephants. Although one of the elephants died, the other, named Tina, remained standing, but with knees beginning to give way. Two family members, Trista and Teresia (Tina's mother), walked to both sides of Tina and leaned in to hold her up. Eventually, Tina grew so weak, she fell to the ground and died. However, Trista and Teresia did not give up but continually tried to lift her. They managed to get Tina into a sitting position, but her body was lifeless and fell to the ground again. As the other elephant family members became more intensely involved in the aid, they tried to put grass into Tina's mouth. Teresia then put her tusks beneath Tina's head and front quarters and proceeded to lift her. As she did so, her right tusk broke completely off, right up to the lip and nerve cavity. The elephants gave up trying to lift Tina but did not leave her; instead, they began to bury her in a shallow grave and throw leaves over her body. They stood over Tina for the night and then began to leave in the morning. The last to leave was Teresia.

Because elephants are so closely knit and highly matriarchal, a family can be devastated by the death of another (especially a matriarch), and some groups never recover their organization. Cynthia Moss has observed a mother, after the death of her calf, walk sluggishly at the back of a family for many days.

Edward Topsell stated in his publication *The History of Four-Footed Beasts* in 1658, "There is no creature among all the Beasts of the world which hath so great and ample demonstration of the power and wisdom of almighty God as the elephant. Elephants are believed to be on par with chimpanzees with regards to their cooperative skills.

ELEPHANT ALTRUISM

Elephants are thought to be highly altruistic animals that even aid other species, including humans, in distress. In India, an elephant was helping locals lift logs by following a truck and placing the logs in pre-dug holes upon instruction from the mahout (elephant trainer). At a certain hole, the elephant refused to lower the log. The mahout came to investigate the hold-up and noticed a dog sleeping in the hole. The elephant only lowered the log when the dog was gone. When an elephant is hurt, other elephants (even if they are unrelated) aid them.

Cynthia Moss has often seen elephants going out of their way to avoid hurting or killing a human, even when it was difficult for them (such as having to walk backwards to avoid a person). Joyce Poole documented an encounter told to her by Colin Francombe on Kuki Gallman's Laikipia Ranch. A ranch herder was out on his own with camels when he came across a family of elephants. The matriarch charged at him and knocked him over with her trunk, breaking one of his legs. In the evening, when he did not return, a search party was sent in a truck to find him. When the party discovered him, he was being guarded by an elephant. The animal charged the truck, so they shot over her and scared her away. The herdsman later told them that when he could not stand up, the elephant used her trunk to lift him under the shade of a tree. She guarded him for the day and would gently touch him with her trunk.

SELF-MEDICATION

Elephants in Africa self-medicate by chewing on the leaves of a tree from the family Boraginaceae, which induces labour. Kenyans also use this tree for the same purpose.

DEATH RITUAL

Elephants are one of the only species of mammals other than *Homo sapiens sapiens* and Neanderthals known to have or have had any recognizable ritual around death. They (elephants) show a keen interest in the bones of their own kind (even unrelated elephants that have died long ago). They are often seen gently investigating the bones with their trunks and feet while remaining very quiet. Sometimes elephants that are completely unrelated to the deceased still visit their graves.

Elephant researcher Martin Meredith recalls in his book an occurrence of a typical elephant death ritual as witnessed by Anthony Hall-Martin, a South African biologist who had studied elephants in Addo, South Africa, for over eight years. The entire family of a dead matriarch, including her young calf, were all gently touching her body with their trunks, trying to lift her. The elephant herd were all rumbling loudly. The calf was observed to be weeping and made sounds that sounded like a scream, but then the entire herd fell incredibly silent. They then began to throw leaves and dirt over the body and broke off tree branches to cover her. They spent the next two days quietly standing over her body. They sometimes had to leave to get water or food, but they would always return.

Occurrences of elephants behaving this way around human beings are common throughout Africa. On many occasions, they have buried dead or sleeping humans or aided them when they were hurt. Meredith also recalls an event told to him by George Adamson, a Kenyan Game Warden, regarding an old Turkana woman who fell asleep under a tree after losing her way home. When she woke up, there was an elephant standing over her, gently touching her. She kept very still because she was very frightened. As other elephants arrived, they began to scream loudly and buried her under branches. She was found the next morning by the local herdsmen, unharmed.

George Adamson also recalls when he shot a bull elephant from a herd that kept breaking into the government gardens of Northern Kenya. George gave the elephant's meat to local Turkana tribesmen and then dragged the rest of the carcass half a mile (800 m) away. That night, the other elephants found the body and took the shoulder blade and leg bone and returned the bones to the exact spot the elephant was killed. Scientists often debate the extent that elephants feel emotion.

PLAY

Joyce Poole on many occasions has observed wild African elephants at play. They apparently do things for their own and others' entertainment. Elephants have been seen sucking up water, holding their trunk high in the air, and then spraying the water like a fountain.

MIMICRY

Recent studies have shown that elephants can also mimic sounds they hear. The discovery was found when Mlaika, an orphaned elephant, would copy the sound of trucks passing by. So far, the only other animals that are thought to mimic sounds are whales, dolphins, bats, primates and birds. Calimero, an African elephant who was 23 years old, also exhibited a unique form of mimicry. He was in a Swiss zoo with some Asian elephants. Asian elephants use chirps that are different from African elephants' deep rumbling noises. Calimero also began to chirp and not make the deep calls that his species normally would.

Kosik, an Indian elephant at Everland Amusement Park, South Korea can imitate up to five Korean words, including *sit*, *no*, *yes* and *lie down*. Kosik produces these human-like sounds by putting his trunk in his mouth and then shaking it while breathing out, similar to how people whistle with their fingers.

Elephants use contact calls to stay in touch with one another when they are out of one another's sight. Female elephants are able to remember and distinguish the contact calls of female family and bond group members from those of females outside of their extended family network. They can also distinguish between the calls of family units depending upon how frequently they came across them.

Joyce Poole, of the Amboseli Elephant Research Project, Kenya, has demonstrated vocal learning and imitation in elephants of sounds made by each other and in the environment. She is beginning to research whether sounds made by elephants have dialects, a trait that is rare in the animal kingdom.

TOOL USE

Elephants show a remarkable ability to use tools, using their trunks like arms. Elephants have been observed digging holes to drink water and then ripping bark from a tree, chewing it into the shape of a ball, filling in the hole and covering over it with sand to avoid evaporation, then later going back to the spot for a drink. They also often use branches to swat flies or scratch themselves. Elephants have also been known to drop very large rocks onto an electric fence either to ruin the fence or to cut off the electricity. Asian elephants in India have been known to break electric fences using logs and clear the surrounding wires using their tusks to provide a safe passageway.

ART AND MUSIC

An elephant painting

Like several other species that are able to produce abstract art, elephants using their trunks to hold brushes create paintings which some have compared to the work of abstract expressionists. However, it is unclear whether the elephants assign any meaning to the paintings that they have created. Elephant art is now commonly featured at zoos, and is shown in museums and galleries around the world. Ruby at the Phoenix Zoo is considered the original elephant art star, and her paintings have sold for as much as $25,000. Ruby chose her own colors and "had a very keen sense of what color, in what sequence, she wanted". The Asian Elephant Art & Conservation Project, an elephant art academy in New York, teaches retired elephants to paint. For paintings that resemble identifiable objects, teachers give the elephants guidance.An example of this was shown in the TV program *Extraordinary Animals*, in which elephants at a camp in Thailand were able to draw portraits with flowers. Although the images were drawn by the elephants, there was always a trainer assisting and guiding the movement.

A popular video showing an elephant painting a picture of another elephant became widespread on internet news and video websites. The website snopes.com, which specializes in debunking urban legends, lists the video as "partly true", in that the elephant produced the brush strokes, but notes that the similarity of the produced paintings is indicative of a learned sequence of strokes rather than a creative effort on the part of the elephant.

It was noted by ancient Romans and Asian elephant handlers (mahouts) that elephants can distinguish melodies. Performing circus elephants commonly follow musical cues and Adam Forepaugh and Barnum & Bailey circuses even featured "elephant bands". German evolutionary biologist Bernard Rensch studied an elephant's ability to distinguish music, and in 1957 published the results in *Scientific American*. Rensch's test elephant could distinguish 12 tones in the music scale and could remember simple melodies. Even though played on varying instruments and at different pitches, timbres and meters, she recognized the tones a year and a half later. These results have been backed up by the Human-Elephant Learning Project which studies elephant intelligence.

An elephant named Shanthi at the National Zoo in Washington D.C. has displayed the ability to play the harmonica and various horn instruments. She reportedly always ends her songs with a crescendo.

Recording group Thai Elephant Orchestra is an ensemble of elephants who improvise music on specially made instruments with minimal interaction from their handlers. The orchestra was co-founded by pachyderm expert Richard Lair, who works at the Thai Elephant Conservation Center in Lampang, and David Sulzer (artist name, Dave Soldier) who studies the role of dopaminergic synapses in memory consolidation, learning, and behavior at Columbia University. According to neurobiologist Aniruddh Patel, the orchestra's star drummer named Pratidah, exhibits musicality, stating: Either when drumming alone or with the orchestra, Pratidah was remarkably steady,. He also noted that she developed a swing-type rhythm pattern when playing with other elephants.

PROBLEM-SOLVING ABILITY

Elephants are able to spend substantial time working on problems. They are able to change their behavior radically to face new challenges, a hallmark of complex intelligence. A 2010 experiment revealed that in order to reach food, "elephants can learn to coordinate with a partner in a task requiring two individuals to simultaneously pull two ends of the same rope to obtain a reward", putting them on an equal footing with chimpanzees in terms of their level of cooperative skills.

In the 1970s, at Marine World Africa, USA, there lived an Asian elephant named Bandula. Bandula worked out how to break open or unlock several of the pieces of equipment used to keep the shackles on her feet secure. The most complex device was a Brummel hook, a device that closes when two opposite points are slid together. Bandula used to fiddle with the hook until it slid apart when it was aligned. Once she had freed herself, she would help the other elephants escape. In Bandula's case and certainly with other captive elephants, there was an element of deception involved during the escapes, such as the animals looking around making sure no one was watching.

In another case, a female elephant worked out how she could unscrew iron rods with an eye hole that was an inch (2.5 cm) thick. She used her trunk to create leverage and then untwisted the bolt.

Ruby, an Asian elephant at Phoenix Zoo would often eavesdrop on conversations keepers would have talking about her. When she heard the word *paint*, she became very excitable. The colors she favored were green, yellow, blue and red. Once, a fire truck came and parked outside her enclosure where a man had just had a heart attack. The lights on the truck were flashing red, white and yellow. When Ruby painted later on in the day, she chose those colors. She also showed a preference for colors that the keepers wore.

Harry Peachey, an elephant trainer, developed a cooperative relationship with an elephant named Koko. Koko would help the keepers out, "prompting" them to encourage him with various commands and words that Koko would learn. Peachey stated that elephants are almost predisposed to cooperate and work with humans as long as they are treated with respect and sensitivity. Koko worked out when his keepers needed a bit of "elephant help" when they were transferring the females of the group to another zoo. When the keepers wanted to transfer a female, they would usually say her name, followed by the word *transfer* (e.g., "Connie transfer"). Koko soon figured out what this meant. If the keepers asked an elephant to transfer and they did not budge, they would say, "Koko, give me a hand." When he heard this, Koko would help. After 27 years of working with elephants, Peachey firmly believes that they can understand the semantics and syntax of some of the words they hear. This is something thought to be very rare in the animal kingdom.

A study by Dr. Naoko Irie of Tokyo University has shown that elephants demonstrate skills at arithmetic. The experiment consist[ed] of dropping varying numbers of apples into two buckets in front of the [Ueno Zoo] elephants and then recording how often they could correctly choose the bucket holding the most fruit. When more than one apple was being dropped into the bucket, this meant that the elephants had to keep running totals in their heads to keep track of the count. The results showed that Seventy-four percent of the time, the animals correctly picked the fullest bucket. An African elephant named Ashya scored the highest with an amazing eighty-seven percent ... Humans in this same contest managed a success rate of just sixty-seven percent. The study was also filmed to ensure its accuracy.

A study on Discovery News found that elephants, during an intelligence test employing food rewards, had found shortcuts that not even the experiment's researchers had thought of.

According to one source, elephants can figure out how to retrieve distant objects that they cannot otherwise reach by using a stick.

APPLYING THE STRING-DRAWING TASK TO ELEPHANTS

In 1956, W. H. Thorpe explained:

The ability to pull up food which is suspended by a thread, the pulled in loop being held by the foot while the bird reaches with its beak for the next pull, is doubtfully inborn and it has been subject to many experiments. The act appears at first sight to be a real and sudden solution of the problem from the start, and thus to qualify for inclusion under ‹insight learning.' Successful performance in this task has been documented in well over ten bird species.

More recently, Bernd Heinrich and Thomas Bugnyar concluded that ravens' "behaviour in accessing meat on a string is not only a product of rapid learning but may involve some understanding of cause–effect relation between string, food and certain body parts. String-pulling behavior has been likewise studied in seven Asian elephants by presenting them with a retractable (bungee) cord. In this setup, the cord is tied to a heavy log a few meters away from the elephant. A sugarcane (a favorite elephant treat) is attached to the cord, and can only be retrieved by repeated, coordinated, action of the trunk and another body part. The results were clearcut:

All seven logging elephants fully mastered the string-drawing sequence within 1-3 experimental sessions. In all cases of retractable rope pulling, the sequence involved pulling by the trunk, and then securing the rope by either foot or mouth. After 2-6 coordinated pulls, while still holding the rope with either mouth or foot, the elephants disengaged the sugarcane from the rope while still using mouth or forefoot as an anchor, and then consumed the sugarcane. All elephants seemed to be flexible about the use of anchor, interchangeably using mouth, foreleg, or both.

SELF-AWARENESS

Asian elephants have joined a small group of animals, including great apes, bottlenose dolphins and eurasian magpies, that exhibit self-awareness. The study was conducted with the Wildlife Conservation Society (WCS) using elephants at the Bronx Zoo in New York. Although many animals respond to a mirror, very few show any evidence that they recognize it is in fact themselves in the mirror reflection.

The Asian elephants in the study also displayed this type of behavior when standing in front of a 2.5-by-2.5-metre (8.2 ft × 8.2 ft) mirror - they inspected the rear and brought food close to the mirror for consumption.

Evidence of elephant self awareness was shown when the elephant Happy repeatedly touched a painted *X* on her head with her trunk, a mark which could only be seen in the mirror. Happy ignored another mark made with colorless paint that was also on her forehead to ensure she was not merely reacting to a smell or feeling.

Frans De Waal, who ran the study, stated, These parallels between humans and elephants suggest a convergent cognitive evolution possibly related to complex society and cooperation.

Self-Awareness and Culling

There has been considerable debate over the issue of culling African elephants in South Africa's Kruger National Park as a means of controlling the population. Some scientists and environmentalists argue that it is "unnecessary and inhumane" to cull them since "elephants resemble humans in a number of ways, not least by having massive brains, social bonds that appear to be empathetic, long gestations, high intelligence, offspring that require an extended period of dependent care, and long life spans." A South African Animal Rights group asked in a statement anticipating the announcement, "How much like us do elephants have to be before killing them becomes murder?"

Others argue that culling is necessary when biodiversity is threatened. However, the protection of biodiversity argument has been questioned by some animal rights advocates who argue that the animal which most greatly threatens and damages biodiversity is humanity, so if we are not willing to cull our own species we cannot morally justify culling another.

TEARS

An elephant, Raju, who had been in captivity for 50 years in the Uttar Pradesh region of India, was freed in a midnight raid by conservationists on July 3–4, 2014. During the release, it was noted that the elephant appeared to be crying, as tears were seen coming from his eyes. The case became widely known, and Raju was called in many headlines and news articles the Crying Elephant.

A baby elephant called Zhuangzhuang was rejected by its mother in 2013. Caretakers thought it was an accident so they treated his injuries and returned him to his mother only for her to reject him again. Caretakers took the elephant away and he cried for 5 hours before the staff could comfort him.

Are elephants highly intelligent?

This section summarizes experiments that cannot be readily reconciled with the view that elephants are highly intelligent. These experiments, in turn, rely on pioneering early work with dogs and cats.

Discrimination Tasks

Thorndike argued that his cats and dogs escaped puzzle boxes through a mindless process of trial and error. Because understanding something as simple as pulling a loop to open a door must occur rapidly or not at all, it should have induced, at some point during the repeated introductions of his animals into the box, a sudden reduction in escape time. The actual, gradual, slope of the time-curve that he did observe suggested to him that his subjects failed to understand the cause-effect relationships between their actions and escape.

In 1957, Altevogt and Rensch reported that a young Asian elephant needed 330 trials, over a period of several days, to consistently choose the reinforced response in her first discrimination task. In an experiment which employed another sense modality, an 8-year-old took 7.5 months to distinguish 12 tones.

Similarly, in discrimination experiments with Asian elephants in the logging camps of Myanmar, only 13 Burmese elephants mastered black/white or large/small visual discrimination tasks, while 7 elephants failed to master this simple task.

Like Thorndike's cats and dogs, the 13 elephants that did master a black/white or large/small discrimination task did so gradually, over several sessions. The slope of the time-curve again suggested the elephants failed to understand the cause-effect relationships between lid removal and food retrieval.

The pre-training that preceded these Myanmar discrimination experiments involved learning to remove a lid from a bucket or to displace a box to uncover a hole in the ground. On average, the 20 elephants taking part in these experiments required 3.4 sessions to gradually master the pre-training task.

Possible Absence of Causal Reasoning

Thorndike's dogs and cats learned to escape a box by, for instance, pulling a loop attached to a string that opened the box. On subsequent introductions to an open box when string-pulling no longer served a purpose, the animal continued to pull a string before getting out of the box. Thorndike concluded that the animal solved the task mechanically, without understanding the causal link between string-pulling and escaping.

A conceptually similar experiment involved pre-training four logging Asian elephants to remove food from a cover-less bucket by inserting their trunk into the bucket. Next, a treat was placed at the bottom of a bucket and, at the same time, the bucket was covered with a lid. The elephants were then trained to obtain the treat by removing the lid. Once this behavioral sequence was established, a treat was placed inside the bucket and, at the same time, the lid was placed on the ground alongside the bucket, so that the lid no longer obstructed access to the food.

As in the Thorndike case, if the elephants understand the nature of the task, they should ignore the lid on the ground and retrieve the reward directly, as they used to do in pre-training sessions before the lid was introduced. But if they fail to understand the causal link between lid removal and food retrieval, they might continue to remove the lid before retrieving the food. The observations accorded entirely with Thorndike's mechanical learning hypothesis.

COGNITIVE BIAS IN ANIMALS

Cognitive bias in animals is a pattern of deviation in judgment, whereby inferences about other animals and situations may be affected by irrelevant information or emotional states. It is sometimes said that animals create their own "subjective social reality" from their perception of the input. In humans, for example, an optimistic or pessimistic bias might affect ones answer to the question "Is the glass half empty or half full?"

To explore cognitive bias, one might train an animal to expect that a positive event follows one stimulus and that a negative event follows another stimulus. For example, on many trials, if the animal presses lever A after a 20 Hz tone it gets a highly desired food, but a press on lever B after a 10 Hz tone yields bland food. The animal is then offered both levers after an intermediate test stimulus e.g. a 15 Hz tone. The hypothesis is that the animal's mood will bias the choice of levers after the test stimulus; if positive, it will tend to choose lever A, if negative it will tend to choose lever B. The hypothesis is tested by manipulating factors that might affect mood – for example, the type of housing the animal is kept in.

Cognitive biases have been shown in a wide range of species including rats, dogs, rhesus macaques, sheep, chicks, starlings and honeybees.

In Rats

In what has been described as a "landmark study", the first study of cognitive bias in animals was conducted with rats. This showed that laboratory rats in unpredictable environments had a more pessimistic attitude than rats in predictable environments.

One study on rats investigated whether changes in light intensity – a short-term manipulation of emotional state – has an effect on cognitive bias. Light intensity was chosen as a treatment because this specifically relates to anxiety-induction. Rats were trained to discriminate between two different locations, in either high ('H') or low ('L') light levels. One location was rewarded with palatable food and the other with aversive food. Rats switched from high to low light levels (putatively the least negative emotional manipulation) ran faster to all three ambiguous locations than rats switched from low to high light levels (putatively the most negative manipulation).

Another study investigated whether chronic social defeat makes rats more pessimistic. To induce chronic psychosocial stress, rats were subjected to daily social defeat in a resident–intruder paradigm for three weeks. This chronic psychosocial stress makes rats more pessimistic.

Using the cognitive bias approach, it has been found that rats which are subjected to either handling or playful, experimenter-administered manual stimulation (tickling) showed different responses to the intermediate stimulus: rats exposed to tickling were more optimistic. The authors stated that they had demonstrated "...for the first time a link between the directly measured positive affective state and decision making under uncertainty in an animal model."

In Pet Dogs

Up to five million pet dogs in the UK, approximately 50% of the population, may perform undesirable separation-related behaviour when left home alone. Dogs were trained to move from a start position to a food bowl. When the bowl was on one side of the room ('positive' location, P) it contained a small quantity of food, and when on the opposite side ('negative' location, N) it was empty. In test trials, the bowl (empty) was placed at one of three ambiguous locations between P and N (near-positive (NP), middle (M), or near-negative (NN). Three test trials were presented at each location. The researchers measured how quickly the dogs moved to the ambiguous locations, fast indicating anticipation of food (an 'optimistic' judgement) or more slowly (a 'pessimistic' judgement. These cognitive bias tests show that dogs which exhibit high levels of separation-related behaviour in a separation test also have a more negative underlying mood.

In Pigs

Domestic pigs do not appear to develop a cognitive bias when kept in different stocking densities. Farmed pigs trained to expect food inside a bowl in one location and not in another, and then tested to show their responses to ambiguous spatial locations. Forty growing pigs were housed in groups of 10 at different density for 8 weeks prior to the start of the test. Tests on three occasions for each pig did not reveal any difference in cognitive bias according to the pig's history of stocking density.

One study shows that restriction of collared peccaries (*Pecari tajacu*) in metabolism pens affects their emotional state and increases feacal glucocorticoid (a stress hormone) metabolite concentrations. The researchers noted that these effects were mitigated by environmental enrichment.

In Honeybees

Honeybees become pessimistic after being shaken

Honeybees ("Apis mellifera carnica") were trained to extend their proboscis to a two-component odour mixture (CS+) predicting a reward (e.g., 1.00 or 2.00 M sucrose) and to withhold their proboscis from another mixture (CS−) predicting either punishment or a less valuable reward (e.g., 0.01 M quinine solution or 0.3 M sucrose). Immediately after training, half of the honeybees were subjected to vigorous shaking for 60 s to simulate the state produced by a predatory attack on a concealed colony. This shaking reduced levels of octopamine, dopamine, and serotonin in the hemolymph of a separate group of honeybees at a time point corresponding to when the cognitive bias tests were performed. In honeybees, octopamine is the local neurotransmitter that functions during reward learning, whereas dopamine mediates the ability to learn to associate odours with quinine punishment. If flies are fed serotonin, they are more aggressive; flies depleted of serotonin still exhibit aggression, but they do so much less frequently.

Within 5 minutes of the shaking, all the trained bees began a sequence of unreinforced test trials with five odour stimuli presented in a random order for each bee: the CS+, the

CS−, and three novel odours composed of ratios intermediate between the two learned mixtures. Shaken honeybees were more likely to withhold their mouthparts from the CS− and from the most similar novel odour. Therefore, agitated honeybees display an increased expectation of bad outcomes similar to a vertebrate-like emotional state. The researchers of the study stated that, Although our results do not allow us to make any claims about the presence of negative subjective feelings in honeybees, they call into question how we identify emotions in any nonhuman animal. It is logically inconsistent to claim that the presence of pessimistic cognitive biases should be taken as confirmation that dogs or rats are anxious but to deny the same conclusion in the case of honeybees.

COGNITIVE ETHOLOGY

Cognitive ethology is a branch of ethology concerned with the influence of conscious awareness and intention on the behaviour of an animal. Donald Griffin, a zoology professor in the United States, set up the foundations for researches in the cognitive awareness of animals within their habitats.

The fusion of cognitive science and classical ethology into **cognitive ethology** "emphasizes observing animals under more-or-less natural conditions, with the objective of understanding the evolution, adaptation (function), causation, and development of the species-specific behavioral repertoire" - (Niko Tinbergen 1963).

According to Jamieson & Bekoff (1993), Tinbergen's four questions about the evolution, adaptation, causation and development of behavior can be applied to the cognitive and mental abilities of animals. Allen & Bekoff (1997, chapter 5) attempt to show how cognitive ethology can take on the central questions of cognitive science, taking as their starting point the four questions described by Barbara Von Eckardt in her 1993 book *What is Cognitive Science?*, generalizing the four questions and adding a fifth. Kingstone, Smilek & Eastwood (2008) suggested that cognitive ethology should include human behavior. They proposed that researchers should firstly study how people behave in their natural, real world environments and then move to the lab. Anthropocentric claims for the ways non-human animals interact in their social and non-social worlds are often used to influence decisions on how the non-human animals can or should be used by humans.

Relation to lAboratory Experimental Psychology

Traditionally, cognitive ethologists have questioned research methods that isolate animals in unnatural surroundings and present them with a limited set of artificial stimuli, arguing that such techniques favor the study of artificial issues that are not relevant to an understanding of the natural behavior of animals. However, many modern researchers favor a judicious combination of field and laboratory methods.

Relation to Ethics

Bekoff, M and Allen, C (1997) "identify three major groups of people (among some of whose members there are blurred distinctions) with different views on cognitive ethology, namely, slayers, skeptics, and proponents." The latter seemingly convergent with animal rights thinking in seeing animal experience as worthy in itself.

Ethicist Peter Singer is an example of a "proponent" in this sense, as is biologist E. O. Wilson who coined the term biophilia to describe the basis of a direct moral cognition, that ‹higher' animals would use to perceive moral implications in the environment directly.

Three Views

According to Marc Bekoff, there are three different views towards whether a science of cognitive ethology is even possible. Slayers deny any possibility of success in cognitive ethology, proponents keep an open mind about animal cognition and the utility of cognitive ethological investigation, and skeptics stand somewhere in between.

REFERENCES

1. http://blogs.scientificamerican.com/brainwaves/2014/02/26/searching-for-the-elephants-genius-inside-the-largest-brain-on-land/
2. Roth, Gerhard; Maxim I. Stamenov; Vittorio Gallese. Is the human brain unique?. Mirror Neurons and the Evolution of Brain and Language. John Benjamins Publishing. pp. 63–76.
3. Goodman, M.; Sterner, K.; Islam, M.; Uddin, M.; Sherwood, C.; Hof, P.; Hou, Z.; Lipovich, L.; Jia, H. (19 November 2009). Phylogenomic analyses reveal convergent patterns of adaptive evolution in elephant and human ancestries(PDF). Proceedings of the National Academy of Sciences. 106 (49): 20824–20829. doi:10.1073/pnas.0911239106. PMC 2791620. PMID 19926857.
4. "Elephants know when they need a helping trunk in a cooperative task". PNAS. Retrieved 2011-03-08.
5. BBC: Elephants know how to co-operate http://news.bbc.co.uk/earth/hi/earth_news/newsid_9417000/9417308.stm
6. Parsell, D.L. (2003-02-21). In Africa, Decoding the Language of Elephants. National Geographic News. Retrieved 2007-10-30.
7. Zimmer, Carl (10 October 2013). Elephants Get the Point of Pointing, Study Shows. New York Times. Retrieved 10 October 2013.
8. Jennifer Viegas (2011). Elephants smart as chimps, dolphins. ABC Science. Retrieved 2011-03-08.
9. Jennifer Viegas (2011). Elephants Outwit Humans During Intelligence Test. Discovery News. Retrieved 2011-03-19.
10. What Makes Dolphins So Smart?. The Ultimate Guide: Dolphins. 1999. Archived from the original on 2012-06-30. Retrieved 2007-10-30.
11. Mind, memory and feelings. Friends Of The Elephant. Retrieved 2007-12-20.
12. Hart, B.L.; L.A. Hart; M. McCoy; C.R. Sarath (November 2001). Cognitive behaviour in Asian elephants: use and modification of branches for fly switching.

Animal Behaviour. Academic Press. 62 (5): 839–847. doi:10.1006/anbe.2001.1815. Retrieved 2007-10-30.

13. Scott, David (2007-10-19). Elephants Really Don't Forget. Daily Express. Archived from the original on 2008-03-12. Retrieved 2007-10-30.
14. Tom, Patrick (2002). The Debate Over Elephant Culling: Is it Ever Morally Justified to Cull Elephants? (PDF). Zambezia. University of Zimbabwe. XXIX (i): 76–81 . Retrieved 2010-08-29.
15. O'Connell, Caitlin (2007). The Elephant's Secret Sense: The Hidden Lives of the Wild Herds of Africa. New York City: Simon & Schuster. pp. 174, 184. ISBN 0-7432-8441-0.
16. Nissani, M. (2008). Elephant Cognition: A Review of Recent Experiments(PDF). Gajah: Journal of the Asian Elephant Specialist Group. 28: 44-52.
17. Elephant brain, Part I: Gross morphology, functions, comparative anatomy, and evolution (PDF). Jeheskel Shoshani, William J. Kupsky b, Gary H. Marchant. Archived from the original (PDF) on 2008-05-09. Retrieved 2007-11-09.
18. Mechanisms of Economic and Social Decision-Making. Allman Labs. Retrieved 2007-11-03.
19. Brain of the African Elephant (Loxodonta africana): Neuroanatomy From Magnetic Resonance Images (PDF). The Anatomical Record Part A: Discoveries in Molecular, Cellular, and Evolutionary Biology. 287: 1117–1127. 2005. Retrieved 2008-01-23.
20. Bekoff, Mark. Do Elephants Cry?: The science is conclusive: animals are emotional beings. Emagazine.
21. Siebert, Charles (October 6, 2006). An Elephant Crack Up?. The New York Times.
22. Shoshani, Jeheskel; Kupsky, William J.; Marchant, Gary H. (30 June 2006). Elephant brain Part I: Gross morphology, functions,comparative anatomy, and evolution. Brain Research Bulletin. 70 (2): 124–157. doi:10.1016/j.brainresbull.2006.03.016. PMID 16782503.
23. Marino, Lori (2004). Cetacean Brain Evolution: Multiplication Generates Complexity (PDF). International Society for Comparative Psychology. The International Society for Comparative Psychology (17): 1–16. Retrieved 2013-08-10.
24. Thinking about Brain Size. Retrieved 2013-08-10.
25. Poole, Joyce (1996). Coming of Age with Elephants. Chicago, Illinois: Trafalgar Square. pp. 131–133, 143–144, 155–157. ISBN 0-340-59179-X.
26. Dolphins Behaviour. Dolphins and Whales Window. Retrieved 2013-03-02.
27. Elephants Brain (PDF). Elsevier. Archived from the original (PDF) on 2008-05-09. Retrieved 2007-10-31.
28. Hakeem, Atiya Y.; Chet. C. Sherwood; Christopher J. Bonar; Camilla Butti; Patrick R. Hof; John M. Allman (December 2009). Von Economo Neurons in the Elephant Brain. The Anatomical Record. 292 (2): 242–248. doi:10.1002/

ar.20829. PMID 19089889.

29. Coghlan, A. (27 November 2006). Whales boast the brain cells that ‹make us human'. New Scientist.

30. Hof, P. R.; Van der Gucht, E. (Jan 2007). Structure of the cerebral cortex of the humpback whale, Megaptera novaeangliae (Cetacea, Mysticeti, Balaenopteridae). Anat Rec. 290 (1): 1–31. doi:10.1002/ar.20407. PMID 17441195.

31. Butti, C; Sherwood, CC; Hakeem, AY; Allman, JM; Hof, PR (2009). Total number and volume of Von Economo neurons in the cerebral cortex of cetaceans.. The Journal of Comparative Neurology. 515 (2): 243–59. doi:10.1002/cne.22055. PMID 19412956.

32. Moss, Cynthia (2001). Elephant Memories: Thirteen Years in the Life of an Elephant Family. Chicago, Illinois: University of Chicago Press. ISBN 0-226-54237-8.

33. Topsell, Edward (1658). The History of Four-Footed Beasts. ISBN 0-415-42695-2.

34. Holdrege, Craig (Spring 2001). Elephantine Intelligence. In Context. The Nature Institute (5). Retrieved 2007-10-30.

35. Linden, Eugene (2002). The Octopus and the Orangutan: More Tales of Animal Intrigue, Intelligence and Ingenuity. New York City: Plume. pp. 16–17, 104–105, 191. ISBN 0-452-28411-2. OCLC 49627740.

36. Meredith, Martin (2004). Elephant Destiny: Biography of an Endangered Species in Africa. Canada: PublicAffairs. pp. 184–186. ISBN 1-58648-233-5.

37. Page, George (1999). The Singing Gorilla: Understanding Animal Intelligence. London, United Kingdom: Headline Book Publishing. pp. 175–177. ISBN 0-7472-7569-6.

38. Elephants Learn Through Copying. BBC News. 2005-03-23. Retrieved 2007-10-31.

39. Elephant Mimics Truck Sounds. Live Science. Retrieved 2007-10-31.

40. Stoeger, Angela; D. Mietchen; S. Oh; S. de Silva; C. Herbst; S. Kwon; W. T. Fitch (2012). An Asian elephant imitates human speech. Current Biology. 22 (22): 2144–2148. doi:10.1016/j.cub.2012.09.022.

41. Kosik, Talking Elephant, Attracts Researchers And Tourists In South Korea (VIDEO). Huffington Post. 2010-10-11. Retrieved 2012-04-28.

42. Size of the Elephant Brain. All Experts. Retrieved 2007-10-31.

43. Flores, Graciela (June 1, 2007). When I see an elephant...paint?. The Scientist. Retrieved 21 March 2013.

44. Mayell, Hillary. Painting Elephants Get Online Gallery. National Geographic News. National Geographic Society. Retrieved 21 March 2013.

45. Mayell, Hillary. Painting Elephants Get Online Gallery. National Geographic News. National Geographic Society. p. 2. Retrieved 21 March 2013.

46. Elephant Painting Rumor. snopes.com. Retrieved 2008-04-03.

47. Elephant Painting. youtube.com. Retrieved 2008-04-03.

48. Scigliano, Eric (December 16, 2000). Think Tank; A Band With a Lot More to Offer Than Talented Trumpeters. New York Times. Arts.
49. Elephant Band : Transcript. PRI : Living on Earth. Air Date: Week of March 2, 2001: World Media Foundation. Retrieved 28 March 2013.
50. Shanthi (elephant) (2012-05-01). Shanthi, the National Zoo's Musical Elephant, Plays the Harmonica!. The National Zoo, Smithsonian Institution, Washington DC: YouTube.
51. Elephants know when they need a helping trunk. New Scientist. Retrieved 2011-03-08.
52. Dubroff, M Dee (August 25, 2010). Are Elephants Smarter than Humans When It Comes to Mental Arithmetic?. Digital Journal. Retrieved 2010-08-29.
53. Chadwick,, D.H. (1992). The Fate of the Elephant. San Francisco: Sierra Club.
54. Thorpe, W.H. (1956). Learning and Instinct in Animals. Cambridge, MA.: Harvard University Press. p. 333.
55. Heinrich, Bernd and Thomas Bugnyar (2005). Testing Problem Solving in Ravens: String-Pulling to Reach Food (PDF). Ethology. 111: 962-976.
56. Nissani, Moti (2008). Inner Life of Elephants, Part 7: Logging Camp / Insight?. youtube.
57. Elephants' Jumbo Mirror Ability. BBC News. 2006-10-31. Retrieved 2007-10-31.
58. Wine, Michael (2007-03-01). Cautious call for elephant cull. The New York Times. Retrieved 2010-08-29.
59. S Africa to allow elephant cull. BBC News. 2008-02-25. Retrieved 2010-08-29.
60. S. Africa elephant culling splits wildlife groups. Associated Press. 2005-11-28. Retrieved 2010-08-29.
61. Brown, Les. 1988. Cruelty to Animals: The Moral Debt.
62. Clarke, Paul A.B. and Andrew Linzey (eds.). 1990. Political Theory and Animal Rights. Winchester, MA: Pluto Press.
63. Edwardi (July 2, 2014). Gajah Ini Menangis Saat Dibebaskan Setelah 50 Tahun Dirantai.
64. Sieczkowski, Cavan (July 7, 2014). Raju The Elephant Cries After Being Rescued Following 50 Years Of Abuse, Chains. The Huffington Post. Retrieved October 19, 2014.
65. Sieczkowski, Cavan (September 3, 2014). Raju The Crying Elephant May Be Forced Back To A Life In Chains. The Huffington Post. Retrieved October 19, 2014.
66. Child, Ben (3 September 2014). The elephant who cried: Raju's story on way to big screen. The Guardian. London: Guardian News and Media. Retrieved October 19, 2014.
67. Bird, Susan (September 8, 2014). Cruel Former Owner of Raju the Elephant Vows to Put Him Back in Chains. Care2. Retrieved October 19, 2014.

68. Baby elephant ‹cries' for five hours after mother tried to stamp him to death shortly after birth. Daily Mail. September 12, 2013. Retrieved November 11, 2015.

69. Baby Elephant Rejected By His Mom Isn't Crying Anymore. The Dodo. November 4, 2015. Retrieved November 11, 2015.

70. Thorndike, E. L. (1911). Animal Intelligence. New York: Macmillan. p. 73.

71. Rensch, B. (1957). The intelligence of elephants. Scientific American. 196: 44-49.

72. Reinert, cited by Alevogt (1990). Proboscideans. In: Grzimek's Encyclopedia of Mammals, Volume. 4. New York: McGraw-Hill. p. 474.

73. Nissani,, M., Hoefler-Nissani, D., Lay, U.T.& Htun, U.W. (2005). Simultaneous visual discrimination in Asian elephants. Journal of the Experimental Analysis of Behavior. 83: 15-29.

74. Nissani, M. (2006). Do Asian elephants apply causal reasoning to tool use tasks?. Journal of Experimental Psychology: Animal Behavior Processes. 31: 91-96.

CHAPTER 5

Emotion in Animals

Chapter Contents

The expression of emotions in animals raises a number of stimulating and challenging questions to which relatively little systematic empirical research has been devoted, especially among free-ranging animals. Popular accounts have raised awareness of animal emotions, especially among nonscientists, and provided scientists with much useful information for further systematic research. Such books have also raised hackles among many scientists for being "too soft"—that is, too anecdotal, misleading, or sloppy. However, Burghardt (1997a), despite finding some areas of concern in Masson and McCarthy's book, wrote: "I predict that in a few years the phenomena described here will be confirmed, qualified, and extended".

INTRODUCTION

Emotions in animals are the subjective feelings and emotions experienced by non-human animals. Emotions may be described as subjective, conscious experiences characterized primarily by psychophysiological expressions, biological reactions, and mental states.

A drawing of a cat by T. W. Wood in Charles Darwin's book *The Expression of the Emotions in Man and Animals*, described as being "in an affectionate frame of mind".

Charles Darwin was one of the first scientists to write about the existence and nature of emotions in animals. His observational and sometimes anecdotal approach has developed into a more robust, hypothesis-driven, scientific approach. General hypotheses relating to correlates between humans and animals also support the claim that animals may feel emotions and that human emotions evolved from the same mechanisms. Several tests, such as cognitive bias tests and learned helplessness models, have been developed. Cognitive biases (feelings of optimism or pessimism) have been shown in a wide range of species including rats, dogs, cats, rhesus macaques, sheep, chicks, starlings, pigs and honeybees.

Some behaviourists claim stimulus–response models provide a sufficient explanation for animal behaviours that have been described as emotional, and that it is unnecessary to postulate that animals are conscious. Other behaviourists further question whether animals feel emotions on the grounds that emotions aren't universal even among humans, that interpretations of animal behaviour are anthropomorphic, and that definitions of emotions lack robustness.

ETYMOLOGY, DEFINITIONS, AND DIFFERENTIATION

The word "emotion" dates back to 1579, when it was adapted from the French word *émouvoir*, which means "to stir up". However, the earliest precursors of the word likely date back to the very origins of language.

Emotions have been described as discrete and consistent responses to internal or external events which have a particular significance for the organism. Emotions are brief in duration and consist of a coordinated set of responses, which may include physiological, behavioural, and neural mechanisms. Emotions have also been described as the result of evolution because they provided good solutions to ancient and recurring problems that faced ancestors.

Laterality

It has been proposed that negative, withdrawal-associated emotions are processed predominantly by the right hemisphere, whereas the left hemisphere is largely responsible for processing positive, approach-related emotions. This has been called the "laterality-valence hypothesis".

Basic and Complex Human Emotions

In humans, a distinction is sometimes made between "basic" and "complex" emotions. Six emotions have been classified as basic: anger, disgust, fear, happiness, sadness and surprise. Complex emotions would include contempt, jealousy and sympathy. However, this distinction is difficult to maintain, and animals are often said to express even the complex emotions.

BACKGROUND

Behaviourist Approach

A squirrel communicating with its pup

Prior to the development of animal sciences such as comparative psychology and ethology, interpretation of animal behaviour tended to favour a minimalistic approach known as behaviourism. This approach refuses to ascribe to an animal a capability beyond the least demanding that would explain a behaviour; anything more than this is seen as unwarranted anthropomorphism. The behaviourist argument is, why should humans postulate consciousness and all its near-human implications in animals to explain some behaviour, if mere stimulus-response is a sufficient explanation to produce the same effects?

The cautious wording of Dixon exemplifies this viewpoint:

Moussaieff Masson and McCarthy describe a similar view (with which they disagree):

"Recent work in the area of ethics and animals suggests that it is philosophically legitimate to ascribe emotions to animals. Furthermore, it is sometimes argued that emotionality is a morally relevant psychological state shared by humans and non humans. What is missing from the philosophical literature that makes reference to emotions in animals is an attempt to clarify and defend some particular account of the nature of emotion, and the role that emotions play in a characterization of human nature. I argue in this paper that some analyses of emotion are more credible than others. Because this is so, the thesis that humans and nonhumans share emotions may well be a more difficult case to make than has been recognized thus far."

"While the study of emotion is a respectable field, those who work in it are usually academic psychologists who confine their studies to human emotions. The standard reference work, The Oxford Companion to Animal Behaviour, advises animal behaviourists that "One is well advised to study the behaviour, rather than attempting to get at any underlying emotion. There is considerable uncertainty and difficulty related to the interpretation and ambiguity of emotion: an animal may make certain movements and sounds, and show certain brain and chemical signals when its body is damaged in a particular way. But does this mean an animal feels—is aware of—pain as we are, or does it merely mean it is programmed to act a certain way with certain stimuli? Similar questions can be asked of any activity an animal (including a human) might undertake, in principle. Many scientists regard all emotion and cognition (in humans and animals) as having a purely mechanistic basis."

Because of the philosophical questions of consciousness and mind that are involved, many scientists have stayed away from examining animal and human emotion, and have instead studied measurable brain functions through neuroscience.

Comparative Approach

In 1903, C. Lloyd Morgan published Morgan's Canon, a specialised form of Occam's razor used in ethology, in which he stated:

"In no case is an animal activity to be interpreted in terms of higher psychological processes,

if it can be fairly interpreted in terms of processes which stand lower in the scale of psychological evolution and development."

Darwin's Approach

Charles Darwin initially planned to include a chapter on emotion in *The Descent of Man* but as his ideas progressed they expanded into a book, *The Expression of the Emotions in Man and Animals*. Darwin proposed that emotions are adaptive and serve a communicative and motivational function, and he stated three principles that are useful in understanding emotional expression: First, *The Principle of Serviceable Habits* takes a Lamarckian stance by suggesting that emotional expressions that are useful will be passed on to the offspring. Second, *The*

Principle of Antithesis suggests that some expressions exist merely because they oppose an expression that is useful. Third, *The Principle of the Direct Action of the Excited Nervous System on the Body* suggests that emotional expression occurs when nervous energy has passed a threshold and needs to be released.

Darwin saw emotional expression as an outward communication of an inner state, and the form of that expression often carries beyond its original adaptive use. For example, Darwin remarks that humans often present their canine teeth when sneering in rage, and he suggests that this means that a human ancestor probably utilized their teeth in aggressive action. A domestic dog's simple tail wag may be used in subtly different ways to convey many meanings as illustrated in Darwin's *The Expression of the Emotions in Man and Animals* published in 1872.

- Examples of tail position indicating different emotions in dogs

"Small dog watching a cat on a table"

"Dog approaching another dog with hostile intentions"

"Dog in a humble and affectionate frame of mind"

FIG. 7.—Half-bred Shepherd Dog in the same state as in Fig. 5. By Mr. A. May.

"Half-bred shepherd dog"

FIG. 8.—The same caressing his master. By Mr. A. May.

"Dog caressing his master"

Anecdotal Approach

Evidence for emotions in animals has been primarily anecdotal, from individuals who interact with pets or captive animals on a regular basis. However, critics of animals having emotions often suggest that anthropomorphism is a motivating factor in the interpretation of the observed behaviours. Much of the debate is caused by the difficulty in defining emotions and the cognitive requirements thought necessary for animals to experience emotions in a similar way to humans. The problem is made more problematic by the difficulties in testing for emotions in animals. What is known about human emotion is almost all related or in relation to human communication.

Scientific Approach

In recent years, the scientific community has become increasingly supportive of the idea of emotions in animals. Scientific research has provided insight into similarities of physiological changes between humans and animals when experiencing emotion.

Much support for animal emotion and its expression results from the notion that feeling emotions doesn't require significant cognitive processes, rather, they could be motivated by the processes to act in an adaptive way, as suggested by Darwin. Recent attempts in studying emotions in animals have led to new constructions in experimental and information gathering. Professor Marian Dawkins suggested that emotions could be studied on a functional or a mechanistic basis. Dawkins suggests that merely mechanistic or functional research will provide the answer on its own, but suggests that a mixture of the two would yield the most significant results.

Functional

Functional approaches rely on understanding what roles emotions play in humans and examining that role in animals. A widely used framework for viewing emotions in a functional context is that described by Oatley and Jenkins who see emotions as having three stages: (i) appraisal in which there is a conscious or unconscious evaluation of an event as relevant to a particular goal. An emotion is positive when that goal is advanced and negative when it is impeded (ii) action readiness where the emotion gives priority to one or a few kinds of action and may give urgency to one so that it can interrupt or compete with others and (iii) physiological changes, facial expression and then behavioural action. The structure, however, may be too broad and could be used to include all the animal kingdom as well as some plants.

Mechanistic

The second approach, mechanistic, requires an examination of the mechanisms that drive emotions and search for similarities in animals.

The mechanistic approach is utilized extensively by Paul, Harding and Mendl. Recognizing the difficulty in studying emotion in non-verbal animals, Paul et al. demonstrate possible ways to better examine this. Observing the mechanisms that function in human emotion expression, Paul et al. suggest that concentration on similar mechanisms in animals can provide clear insights into the animal experience. They noted that in humans, cognitive biases vary according to emotional state and suggested this as a possible starting point to examine animal emotion. They propose that researchers may be able to use controlled stimuli which have a particular meaning to trained animals to induce particular emotions in these animals and assess which types of basic emotions animals can experience.

Cognitive Bias Test

A cognitive bias is a pattern of deviation in judgment, whereby inferences about other animals and situations may be drawn in an illogical fashion. Individuals create their own "subjective social reality" from their perception of the input. It refers to the question "Is the glass half empty or half full?", used as an indicator of optimism or pessimism. To test this in animals, an individual is trained to anticipate that stimulus A, e.g. a 20 Hz tone, precedes a positive event, e.g. highly desired food is delivered when a lever is pressed by the animal. The same individual is trained to anticipate that stimulus B, e.g. a 10 Hz tone, precedes a negative event, e.g. bland food is delivered when the animal presses a lever. The animal is then tested by being played an intermediate stimulus C, e.g. a 15 Hz tone, and observing whether the animal presses the lever associated with the positive or negative reward, thereby indicating whether the animal is in a positive or negative mood. This might be influenced by, for example, the type of housing the animal is kept in.

Using this approach, it has been found that rats which are subjected to either handling or playful, experimenter-administered manual stimulation (tickling) showed different responses to the intermediate stimulus: rats exposed to tickling were more optimistic. The authors stated that they had demonstrated "...for the first time a link between the directly measured positive affective state and decision making under uncertainty in an animal model."

Cognitive biases have been shown in a wide range of species including rats, dogs, rhesus macaques, sheep, chicks, starlings and honeybees.

Self-Medication with Psychoactive Drugs

Humans can suffer from a range of emotional or mood disorders such as depression, anxiety, fear and panic. To treat these disorders, scientists have developed a range of psychoactive drugs such as anxiolytics. Many of these drugs are developed and tested by using a range of laboratory species. It is inconsistent to argue that these drugs are effective in treating human emotions whilst denying the experience of these emotions in the laboratory animals on which they have been developed and tested.

Standard laboratory cages prevent mice from performing several natural behaviours for which they are highly motivated. As a consequence, laboratory mice sometimes develop abnormal behaviours indicative of emotional disorders such as depression and anxiety. To improve welfare, these cages are sometimes enriched with items such as nesting material, shelters and running wheels. Sherwin and Ollson tested whether such enrichment influenced the consumption of Midazolam, a drug widely used to treat anxiety in humans. Mice in standard cages, standard cages but with unpredictable husbandry, or enriched cages, were given a choice of drinking either non-drugged water or a solution of the Midazolam. Mice in the standard and unpredictable cages drank a greater proportion of the anxiolytic solution than mice from enriched cages, indicating that mice from the standard and unpredictable laboratory caging may have been experiencing greater anxiety than mice from the enriched cages.

Spindle Neurons

Spindle neurons are specialised cells found in three very restricted regions of the human brain – the anterior cingulate cortex, the frontoinsular cortex and the dorsolateral prefrontal cortex. The first two of these areas regulate emotional functions such as empathy, speech, intuition, rapid "gut reactions" and social organization in humans. Spindle neurons are also found in the brains of humpback whales, fin whales, killer whales, sperm whales, bottlenose dolphin, Risso's dolphin, beluga whales, and the African and Asian elephants.

Whales have spindle cells in greater numbers and are maintained for twice as long as humans. The exact function of spindle cells in whale brains is still not understood, but Hof and Van Der Gucht believe that they act as some sort of "high-speed connections that fast-track information to and from other parts of the cortex". They compared them to express trains that bypass unnecessary connections, enabling organisms to instantly process and act on emotional cues during complex social interactions. However, Hof and Van Der Gucht clarify that they do not know the nature of such feelings in these animals and that we cannot just apply what we see in great apes or ourselves to whales. They believe that more work is needed to know whether emotions are the same for humans and whales.

Vocalisations

From at least the time of Darwin, it has been known that chimpanzees and other great apes also perform a laugh-like vocalization.

Research with rats has revealed that under particular conditions, they emit 50-kHz ultrasonic vocalisations (USV) which have been postulated to reflect a positive affective state (emotion) analogous to primitive human joy; these calls have been termed "laughter". The 50 kHz USVs in rats are uniquely elevated by hedonic stimuli—such as tickling, rewarding electrical brain stimulation, amphetamine injections, mating, play, and aggression—and are suppressed by aversive stimuli. Of all manipulations that elicit 50 kHz chirps in rats, tickling by humans elicits the highest rate of these calls.

Some vocalizations of domestic cats, such as purring, are well-known to be produced in situations of positive valence, such as mother kitten interactions, contacts with familiar partner, or during tactile stimulation with inanimate objects as when rolling and rubbing. Therefore, purring can be generally considered as an indicator of "pleasure" in cats.

Low pitched bleating in sheep has been associated with some positive-valence situations, as they are produced by males as an estrus female is approaching or by lactating mothers while licking and nursing their lambs.

CRITICISM

The argument that animals experience emotions is sometimes rejected due to a lack of evidence, and those who don't believe in the idea of animal intelligence, often argue that anthropomorphism plays a role in individuals' perspectives. Those who reject that animals have the capacity to experience emotion do so mainly by referring to inconsistencies in studies that have endorsed the belief emotions exist. Having no direct means to communicate emotion beyond behavioral response interpretation, the difficulty of providing an account of emotion in animals relies heavily on interpretive experimentation, that relies on results from human subjects.

Some who oppose the concept of animal emotions suggest that emotions aren't universal, including in humans. If emotions are not universal, this indicates that there is not a phylogenetic relationship between human and non-human emotion. The relationship drawn by proponents of animal emotion, then, would be merely a suggestion of mechanistic features that promote adaptivity, but lack the complexity of human emotional constructs. Thus, a social life-style may play a role in basic emotions developing into complex emotions.

Darwin concluded, through a survey, that humans share universal emotive expressions and suggested that animals likely share in these to some degree. Darwin's results have been criticised by individuals that suggest misinterpretations. Social constructivists disregard the concept that emotions are universal.Others hold an intermediate stance, suggesting that basic emotional expressions and emotion are universal but the intricacies are developed culturally. A study by Elfenbein and Ambady indicated that individuals within a particular culture are better at recognising other cultural members' emotions.

EXAMPLES

Primates

Primates, in particular great apes, are candidates for being able to experience empathy and theory of mind. Great apes have complex social systems; young apes and their mothers have strong bonds of attachment and when a baby chimpanzee or gorilla dies, the mother will not uncommonly carry the body around for several days. Jane Goodall has described chimpanzees as exhibiting mournful behavior. Koko, a gorilla trained to use sign language, was reported to have expressed vocalisations indicating sadness after the death of her pet cat, All Ball.

Beyond such anecdotal evidence, support for empathetic reactions has come from experimental studies of rhesus macaques. Macaques refused to pull a chain that delivered food to themselves if doing so also caused a companion to receive an electric shock. This inhibition of hurting another conspecific was more pronounced between familiar than unfamiliar macaques, a finding similar to that of empathy in humans.

Furthermore, there has been research on consolation behavior in chimpanzees. De Waal and Aureli found that third-party contacts attempt to relieve the distress of contact participants by consoling (e.g. making contact, embracing, grooming) recipients of aggression, especially those that have experienced more intense aggression. Researchers were unable to replicate these results using the same observation protocol in studies of monkeys, demonstrating a possible difference in empathy between monkeys and apes.

Other studies have examined emotional processing in the great apes. Specifically, chimpanzees were shown video clips of emotionally charged scenes, such as a detested veterinary procedure or a favorite food, and then were required to match these scenes with one of two species-specific facial expressions: "happy" (a play-face) or "sad" (a teeth-baring expression seen in frustration or after defeat). The chimpanzees correctly matched the clips to the facial expressions that shared their meaning, demonstrating that they understand the emotional significance of their facial expressions. Measures of peripheral skin temperature also indicated that the video clips emotionally affected the chimpanzees.

Rodents

In 1998, Jaak Panksepp proposed that all mammalian species are equipped with brains capable of generating emotional experiences. Subsequent work examined studies on rodents to provide foundational support for this claim. One of these studies examined whether rats would work to alleviate the distress of a conspecific. Rats were trained to press a lever to avoid the delivery of an electric shock, signaled by a visual cue, to a conspecific. They were then tested in a situation in which either a conspecific or a Styrofoam block was hoisted into the air and could be lowered by pressing a lever. Rats that had previous experience with conspecific distress demonstrated greater than ten-fold more responses to lower a distressed conspecific compared to rats in the control group, while those who had never experienced conspecific distress expressed greater than three-fold more responses to lower a distressed conspecific relative to the control group. This suggests that rats will actively work to reduce the distress of a conspecific, a phenomenon related to empathy. Comparable results have also been found in similar experiments designed for monkeys.

Langford et al. examined empathy in rodents using an approach based in neuroscience. They reported that (1) if two mice experienced pain together, they expressed greater levels of pain-related behavior than if pain was experienced individually, (2) if experiencing different levels of pain together, the behavior of each mouse was modulated by the level of pain experienced by its social partner, and (3) sensitivity to a noxious stimulus was experienced to the same degree by the mouse observing a conspecific in pain as it was by the mouse directly experiencing the painful stimulus. The authors suggest this responsiveness to the pain of others demonstrated by mice is indicative of emotional contagion, a phenomenon associated with empathy, which has also been reported in pigs. One behaviour associated with fear in rats is freezing. If female rats experience electric shocks to the feet and then witness another rat experiencing similar footshocks, they freeze more than females without any experience of the shocks. This suggests empathy in experienced rats witnessing another individual being shocked. Furthermore, the demonstrator's behaviour was changed by the behaviour of the witness; demonstrators froze more following footshocks if their witness froze more creating an empathy loop.

Several studies have also shown rodents can respond to a conditioned stimulus that has been associated with the distress of a conspecific, as if it were paired with the direct experience of an unconditioned stimulus. These studies suggest that rodents are capable of shared affect, a concept critical to empathy.

Horses

Although not direct evidence that horses experience emotions, a 2016 study showed that domestic horses react differently to seeing photographs of positive (happy) or negative (angry) human facial expressions. When viewing angry faces, horses look more with their left eye which is associated with perceiving negative stimuli. Their heart rate also increases more quickly and they show more stress-related behaviours.

Birds

Marc Bekoff reported accounts of animal behaviour which he believed was evidence of animals being able to experience emotions in his book *The Emotional Lives of Animals*. The following is an excerpt from his book:

"A few years ago my friend Rod and I were riding our bicycles around Boulder, Colorado, when we witnessed a very interesting encounter among five magpies. Magpies are corvids, a very intelligent family of birds. One magpie had obviously been hit by a car and was laying dead on the side of the road. The four other magpies were standing around him. One approached the corpse, gently pecked at it-just as an elephant noses the carcass of another elephant- and stepped back. Another magpie did the same thing. Next, one of the magpies flew off, brought back some grass, and laid it by the corpse. Another magpie did the same. Then, all four magpies stood vigil for a few seconds and one by one flew off."

Bystander affiliation is believed to represent an expression of empathy in which the bystander tries to console a conflict victim and alleviate their distress. There is evidence for bystander affiliation in ravens (e.g. contact sitting, preening, or beak-to-beak or beak-to-body touching) and also for solicited bystander affiliation, in which there is post-conflict affiliation from the victim to the bystander. This indicates that ravens may be sensitive to the emotions of others, however, relationship value plays an important role in the prevalence and function of these post-conflict interactions.

The capacity of domestic hens to experience empathy has been studied. Mother hens show one of the essential underpinning attributes of empathy: the ability to be affected by, and share, the emotional state of their distressed chicks. However, evidence for empathy between familiar adult hens has not yet been found.

Dogs

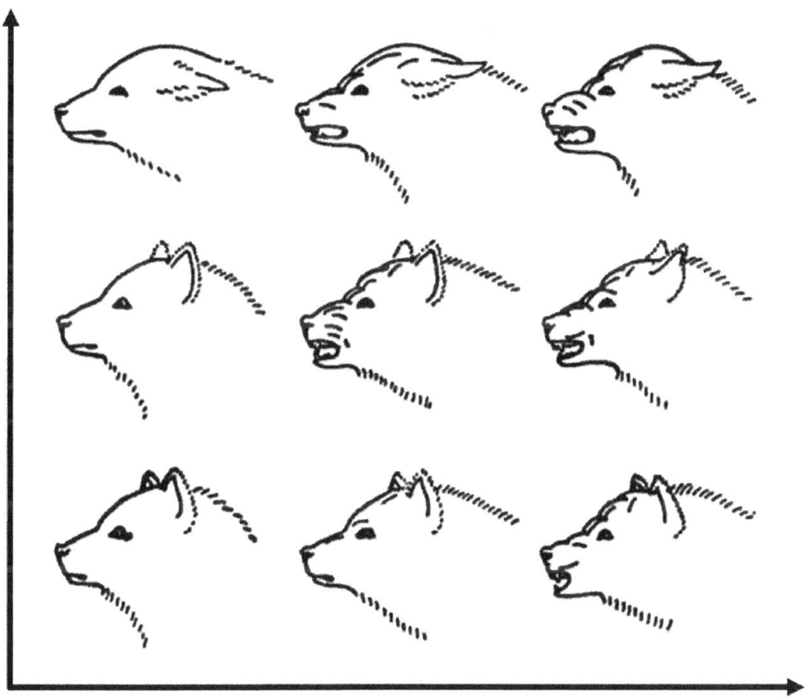

A drawing by Konrad Lorenz showing facial expressions of a dog

Some research indicates that domestic dogs may experience negative emotions in a similar manner to humans, including the equivalent of certain chronic and acute psychological conditions.

Much of this is from studies by Martin Seligman on the theory of learned helplessness as an extension of his interest in depression:

A dog that had earlier been repeatedly conditioned to associate an audible stimulus with inescapable electric shocks did not subsequently try to escape the electric shocks after the warning was presented, even though all the dog would have had to do is jump over a low divider within ten seconds. The dog didn't even try to avoid the "aversive stimulus"; it had previously "learned" that nothing it did would reduce the probability of it receiving a shock. A follow-up experiment involved three dogs affixed in harnesses, including one that received shocks of identical intensity and duration to the others, but the lever which would otherwise have allowed the dog a degree of control was left disconnected and didn't do anything. The first two dogs quickly recovered from the experience, but the third dog suffered chronic symptoms of clinical depression as a result of this perceived helplessness.

A further series of experiments showed that, similar to humans, under conditions of long-term intense psychological stress, around one third of dogs do not develop learned helplessness or long term depression. Instead these animals somehow managed to find a way to handle the unpleasant situation in spite of their past experience. The corresponding characteristic in humans has been found to correlate highly with an explanatory style and optimistic attitude that views the situation as *other than* personal, pervasive, or permanent.

Since these studies, symptoms analogous to clinical depression, neurosis, and other psychological conditions have also been accepted as being within the scope of emotion in domestic dogs.

Psychology research has shown that when humans gaze at the face of another human, the gaze is not symmetrical; the gaze instinctively moves to the right side of the face to obtain information about their emotions and state. Research at the University of Lincoln shows that dogs share this instinct when meeting a human, and only when meeting a human (i.e., not other animals or other dogs). They are the only non-primate species known to share this instinct.

The existence and nature of personality traits in dogs have been studied (15,329 dogs of 164 different breeds). Five consistent and stable "narrow traits" were identified, described as playfulness, curiosity/fearlessness, chase-proneness, sociability and aggressiveness. A further higher order axis for shyness–boldness was also identified.

Dogs presented with images of either human or dog faces with different emotional states (happy/playful or angry/aggressive) paired with a single vocalization (voices or barks) from the same individual with either a positive or negative emotional state or brown noise. Dogs look longer at the face whose expression is congruent to the emotional state of the vocalization, for both other dogs and humans. This is an ability previously known only in humans.

Some researchers are beginning to investigate the question of whether dogs have emotions with the help of magnetic resonance imaging.

Cats

It has been postulated that domestic cats can learn to manipulate their owners through vocalizations that are similar to the cries of human babies. Some cats learn to add a purr to the vocalization, which makes it less harmonious and more dissonant to humans, and therefore harder to ignore. Individual cats learn to make these vocalizations through trial-

and-error; when a particular vocalization elicits a positive response from a human, the probability increases that the cat will use that vocalization in the future.

Growling can be an expression of annoyance or fear, similar to humans. When annoyed or angry, a cat wriggles and thumps its tail much more vigorously than when in a contented state. In larger felids such as lions, what appears to be irritating to them varies between individuals. A male lion may let his cubs play with his mane or tail, or he may hiss and hit them with his paws. Domestic male cats also have variable attitudes towards their family members, for example, older male siblings tend not to go near younger or new siblings and may even show hostility toward them.

Honeybees

Honeybees ("Apis mellifera carnica") were trained to extend their proboscis to a two-component odour mixture (CS+) predicting a reward (e.g., 1.00 or 2.00 M sucrose) and to withhold their proboscis from another mixture (CS−) predicting either punishment or a less valuable reward (e.g., 0.01 M quinine solution or 0.3 M sucrose). Immediately after training, half of the honeybees were subjected to vigorous shaking for 60 s to simulate the state produced by a predatory attack on a concealed colony. This shaking reduced levels of octopamine, dopamine, and serotonin in the hemolymph of a separate group of honeybees at a time point corresponding to when the cognitive bias tests were performed. In honeybees, octopamine is the local neurotransmitter that functions during reward learning, whereas dopamine mediates the ability to learn to associate odours with quinine punishment. If flies are fed serotonin, they are more aggressive; flies depleted of serotonin still exhibit aggression, but they do so much less frequently.

Within 5 minutes of the shaking, all the trained bees began a sequence of unreinforced test trials with five odour stimuli presented in a random order for each bee: the CS+, the CS−, and three novel odours composed of ratios intermediate between the two learned mixtures. Shaken honeybees were more likely to withhold their mouthparts from the CS− and from the most similar novel odour. Therefore, agitated honeybees display an increased expectation of bad outcomes similar to a vertebrate-like emotional state. The researchers of the study stated that, "Although our results do not allow us to make any claims about the presence of negative subjective feelings in honeybees, they call into question how we identify emotions in any non-human animal. It is logically inconsistent to claim that the presence of pessimistic cognitive biases should be taken as confirmation that dogs or rats are anxious but to deny the same conclusion in the case of honeybees."

Crayfish

Crayfish naturally explore new environments but display a general preference for dark places. A 2014 study on the freshwater crayfish *Procambarus clarkii* tested their responses in a fear paradigm, the elevated plus maze in which animals choose to walk on an elevated cross which offers both aversive and preferable conditions (in this case, two arms were lit and two were dark). Crayfish which experienced an electric shock displayed enhanced fearfulness or anxiety as demonstrated by their preference for the dark arms more than the light. Furthermore, shocked crayfish had relatively higher brain serotonin concentrations coupled with elevated blood glucose, which indicates a stress response. Moreover, the crayfish calmed down when they were injected with the benzodiazepine anxiolytic, chlordiazepoxide, used to treat anxiety in humans, and they entered the dark as normal. The authors of the study concluded "...stress-induced avoidance behavior in crayfish exhibits striking homologies with vertebrate anxiety."

A follow-up study using the same species confirmed the anxiolytic effect of chlordiazepoxide, but moreover, the intensity of the anxiety-like behaviour was dependent on the intensity of the electric shock until reaching a plateau. Such a quantitative relationship between stress and anxiety is also a very common feature of human and vertebrate anxiety.

EMPATHY

Empathy is the capacity to understand or feel what another person is experiencing from within the other being's frame of reference, i.e., the capacity to place oneself in another's position. Empathy is seeing with the eyes of another, listening with the ears of another and feelings with the heart of another. There are many definitions for empathy which encompass a broad range of emotional states. Types of empathy include cognitive empathy, emotional empathy, and somatic empathy.

Etymology

The English word *empathy* is derived from the Ancient Greek word εμπάθεια (*empatheia*, meaning "physical affection or passion"). This, in turn, comes from εν (*en*, "in, at") and πάθος (*pathos*, "passion" or "suffering"). The term was adapted by Hermann Lotze and Robert Vischer to create the German word *Einfühlung* ("feeling into"), which was translated by Edward B. Titchener into the English term empathy.

Alexithymia is a word used to describe a deficiency in understanding, processing or describing emotions *in oneself* as opposed to in others. This term comes from the negation of two Ancient Greek words: λέξις (*lexis*, meaning "speech") and *θυμός* (*thymos*, meaning "the soul, as the seat of emotion, feeling and thought"). This combined word is then modified by an alpha privative (prefix *a*-; that is, "not"). Thus alexithymia literally means "not having speech for your emotions".

Definitions

Empathy has many different definitions that encompass a broad range of emotional states, including caring for other people and having a desire to help them; experiencing emotions that match another person's emotions; discerning what another person is thinking or feeling; and making less distinct the differences between the self and the other.

It also is the ability to feel and share another person's emotions. Some believe that empathy involves the ability to match another's emotions, while others believe that empathy involves being tenderhearted toward another person.

Martin Hoffman is a Psychologist who studied the development of empathy. According to Hoffman everyone is born with the capability of feeling empathy.

Compassion and sympathy are two terms that many associate with empathy, but all three of these terms are unique. Even so, definitions vary, contributing to the challenge of defining empathy. Compassion is often defined as an emotion we feel when others are in need, which motivates us to help them. Sympathy is a feeling of care and understanding for someone in need. It can also be understood as having the separateness of defining oneself and another blur. Some include in sympathy also an empathic concern, a feeling of concern for another, in which some scholars include the wish to see them better off or happier.

Empathy is distinct also from pity and emotional contagion. Pity is feeling that another is in trouble and in need of help as they cannot fix their problems themselves, often described as "feeling sorry" for someone. Emotional contagion is when a person (especially an infant or a member of a mob) imitatively "catches" the emotions that others are showing without necessarily recognizing this is happening.

Since empathy involves understanding the emotional states of other people, the way it is characterized is derivative of the way emotions themselves are characterized. If, for example, emotions are taken to be centrally characterized by bodily feelings, then grasping the bodily feelings of another will be central to empathy. On the other hand, if emotions are more centrally characterized by a combination of beliefs and desires, then grasping these beliefs and desires will be more essential to empathy. The ability to imagine oneself as another person is a sophisticated imaginative process. However, the basic capacity to recognize emotions is probably innate and may be achieved unconsciously. Yet it can be trained and achieved with various degrees of intensity or accuracy.

Empathy necessarily has a "more or less" quality. The paradigm case of an empathic interaction, however, involves a person communicating an accurate recognition of the significance of another person's ongoing intentional actions, associated emotional states, and personal characteristics in a manner that the recognized person can tolerate. Recognitions that are both accurate and tolerable are central features of empathy.

The human capacity to recognize the bodily feelings of another is related to one's imitative capacities, and seems to be grounded in an innate capacity to associate the bodily movements and facial expressions one sees in another with the proprioceptive feelings of producing those corresponding movements or expressions oneself. Humans seem to make the same immediate connection between the tone of voice and other vocal expressions and inner feeling.

In the field of positive psychology, empathy has also been compared with altruism and egotism. Altruism is behavior that is aimed at benefitting another person, while egotism is a behavior that is acted out for personal gain. Sometimes, when someone is feeling empathetic towards another, acts of altruism occur. However, many question whether or not these acts of altruism are motivated by egotistical gains. According to positive psychologists, people can be adequately moved by their empathies to be altruistic.

Gender Differences

The literature commonly indicates that females tend to be more empathic than males. On average, female subjects score higher than males on the Empathy Quotient (EQ), while males tend to score higher on the Systemizing Quotient (SQ). Both males and females with autistic spectrum disorders usually score lower on the EQ and SQ (see below for more detail on autism and empathy). However, a series of studies, using a variety of neurophysiological measures, including MEG, spinal reflex excitability, electroencephalography and N400 paradigm have documented the presence of an overall gender difference in the human mirror neuron system, with female participants tending to exhibit stronger motor resonance than male participants. In addition, these aforementioned studies found that female participants tended to score higher on empathy self-report dispositional measures and that these measures positively correlated with the physiological response. Other studies show no significant difference, and suggest gender differences are the result of motivational differences.

A review published in the journal *Neuropsychologia* found that women tended to be better at recognizing facial effects, expression processing and emotions in general. Men only tended to be better at recognizing specific behavior which includes anger, aggression and threatening cues. A 2006 meta-analysis by researcher Rena A Kirkland in the journal *North American Journal of Psychology* found significant sex differences favoring females in "Reading of the mind" test. "Reading of the mind" test is an advanced ability measure of cognitive empathy in which Kirkland's analysis involved 259 studies across 10 countries. Another 2014 meta-analysis in the journal of *Cognition and Emotion*, found overall female advantage in non-verbal emotional recognition across 215 samples.

Using fMRI, neuroscientist Tania Singer showed that empathy-related neural responses tended to be significantly lower in males when observing an "unfair" person experiencing pain. An analysis from the journal of *Neuroscience & Biobehavioral Reviews* also found that, overall, there are sex differences in empathy from birth, growing larger with age and which remains consistent and stable across lifespan. Females, on average, were found to have higher empathy than males, while children with higher empathy regardless of gender continue to be higher in empathy throughout development. Further analysis of brain tools such as event related potentials found that females who saw human suffering tended to have higher ERP waveforms than males. Another investigation with similar brain tools such as N400 amplitudes found, on average, higher N400 in females in response to social situations which positively correlated with self-reported empathy.Structural fMRI studies also found females to have larger grey matter volumes in posterior inferior frontal and anterior inferior parietal cortex areas which are correlated with mirror neurons in fMRI literature. Females also tended to have a stronger link between emotional and cognitive empathy. The researchers found that the stability of these sex differences in development are unlikely to be explained by any environment influences but rather might have some roots in human evolution and inheritance.

Throughout prehistory, females were the primary nurturers and caretakers of children, so this might have led to an evolved neurological adaptation for women to be more aware and responsive to non-verbal expressions. According to the *Primary Caretaker Hypothesis*, prehistoric males did not have the same selective pressure as primary caretakers; so therefore this might explain modern day sex differences in emotion recognition and empathy.

Major Theories and Empirical Findings

Research investigating the social response to natural disasters looked at the characteristics associated with individuals who help victims. Researchers found that cognitive empathy, rather than emotional empathy, predicted helping behavior towards victims. Others have posited that taking on the perspectives of others (cognitive empathy) allows these individuals to better empathize with victims without as much discomfort, whereas sharing the emotions of the victims (emotional empathy) can cause emotional distress, helplessness, victim-blaming, and ultimately can lead to avoidance rather than helping.

Yet, despite this evidence for empathy-induced altruistic motivation, egoistic explanations may still be possible. For example, one alternative explanation for the problem-specific helping pattern may be that the sequence of events in the same problem condition first made subjects sad when they empathized with the problem and then maintained or enhanced subjects' sadness when they were later exposed to the same plight. Consequently, the negative state relief model would predict substantial helping among imagine-set subjects in the same condition, which is what occurred. An intriguing question arises from such findings concerning whether it is possible to have mixed motivations for helping. If this is the case, then simultaneous egoistic and altruistic motivations would occur. This would allow for a stronger sadness-based motivation to obscure the effects of an empathic concern-based altruistic motivation. The observed study would then have sadness as less intense than more salient altruistic motivation. Consequently, relative strengths of different emotional reactions, systematically related to the need situation, may moderate the predominance of egoistic or altruistic motivation (Dovidio, 1990). But it has been shown that researchers in this area who have used very similar procedures sometimes obtain apparently contradictory results. Superficial procedural differences such as precisely when a manipulation is introduced could also lead to divergent results and conclusions. It is therefore vital for any future research to move toward even greater standardization of measurement. Thus, an important step in solving the current theoretical debate concerning the existence of altruism may involve reaching common methodological ground.

Contemporary neuroscience has allowed us to understand the neural basis of the human mind's ability to understand and process emotion. Studies today enable us to see the activation of mirror neurons and attempt to explain the basic processes of empathy. By isolating these mirror neurons and measuring the neural basis for human mind reading and emotion sharing abilities, science has come one step closer to finding the reason for reactions like empathy. Neuroscientists have already discovered that people scoring high on empathy tests have especially busy mirror neuron systems in their brains (Dr. Christian Keysers).Empathy is a spontaneous sharing of affect, provoked by witnessing and sympathizing with another's emotional state. In a way we mirror or mimic the emotional response that we would expect to feel in that condition or context, much like sympathy. Unlike personal distress, empathy is not characterized by aversion to another's emotional response. Additionally, empathizing with someone requires a distinctly sympathetic reaction where personal distress demands avoidance of distressing matters. This distinction is vital because empathy is associated with the moral emotion sympathy, or empathetic concern, and consequently also prosocial or altruistic action. Empathy leads to sympathy by definition unlike the over-aroused emotional response that turns into personal distress and causes a turning-away from another's distress.

In empathy we feel what we believe are the emotions of another, which makes it both affective and cognitive by most psychologists. In this sense, arousal and empathy promote prosocial behavior as we accommodate each other to feel similar emotions. For social beings, negotiating interpersonal decisions is as important to survival as being able navigate the physical landscape. Emotions motivate individual behavior that aids in solving communal challenges as well as guiding group decisions about social exchange. Additionally, recent research has shown individuals who report regular experiences of gratitude engage more frequently in prosocial behaviors. Positive emotions like empathy or gratitude are linked to a more positive continual state and these people are far more likely to help others than those not experiencing a positive emotional state. Thus, empathy's influence extends beyond relating to other's emotions, it correlates with an increased positive state and likeliness to aid others. Measures of empathy show that mirror neurons are activated during arousal of sympathetic responses and prolonged activation shows increased probability to help others.

Another growing focus of investigation is how empathy manifests in education between teachers and learners. Although there is general agreement that empathy is essential in educational settings, research has found that it is difficult to develop empathy in trainee teachers. According to one theory, there are seven components involved in the effectiveness of intercultural communication; empathy was found to be one of the seven. This theory also states that empathy is learnable. However, research also shows that it is more difficult to empathize when there are differences between people including status, culture, religion, language, skin colour, gender, age and so on.

In order to achieve intercultural empathy, psychologists have employed empathy training. One study hypothesized that empathy training would increase the measured level of relational empathy among the individuals in the experimental group when compared to the control group. The study also hypothesized that empathy training would increase communication among the experimental group, and that perceived satisfaction with group dialogue would also increase among the experimental group. To test this, the experimenters used the Hogan Empathy Scale, the Barrett-Lennard Relationship Inventory, and questionnaires. Using these measures, the study found that empathy training was not successful in increasing relational empathy. Also, communication and satisfaction among groups did not increase as a result of the empathy training. While there didn't seem to be a clear relationship between empathy and relational empathy training, the study did report that "relational empathy training appeared to foster greater expectations for a deep dialogic process resulting in treatment differences in perceived depth of communication".

The environment has been another interesting topic of study. Many theorize that environmental factors, such as parenting style and relationships, play a significant role in the development of empathy in children. Empathy promotes pro social relationships, helps mediate aggression, and allows us to relate to others, all of which make empathy an important emotion among children.

A study done by Caroline Tisot looked at how a variety of environmental factors affected the development of empathy in young children. Parenting style, parent empathy, and prior social experiences were looked at. The children participating in the study were asked to complete an effective empathy measure, while the children's parents completed the Parenting Practices Questionnaire, which assesses parenting style, and the Balanced Emotional Empathy scale.

This study found that a few parenting practices – as opposed to parenting style as a whole – contributed to the development of empathy in children. These practices include encouraging the child to imagine the perspectives of others and teaching the child to reflect on his or her own feelings. The results also show that the development of empathy varied based on the gender of the child and parent. Paternal warmth was found to be significantly important, and was positively related to empathy within children, especially in boys. Interestingly, however, maternal warmth was negatively related to empathy within children, especially in girls.

It has also been found that empathy can be disrupted due to trauma in the brain such as a stroke. In most cases empathy is usually impaired if a lesion or stroke occurs on the right side of the brain. In addition to this it has been found that damage to the frontal lobe, which is primarily responsible for emotional regulation, can impact profoundly on a person's capacity to experience empathy toward another individual. People who have suffered from an acquired brain injury also show lower levels of empathy according to previous studies. In fact, more than 50% of people who suffer from a traumatic brain injury self-report a deficit in their empathic capacity. Again, linking this back to the early developmental stages of emotion, if emotional growth has been stunted at an early age due to various factors, empathy will struggle to infest itself in that individuals mind-set as a natural feeling, as they themselves will struggle to come to terms with their own thoughts and emotions. This is again suggestive of the fact that understanding your own emotions is key in being able to identify with another individual's emotional state.

Applications

The empathy-altruism relationship also has broad for whom empathy is felt at the expense of other potential pro-social goals, thus inducing a type of bias. Researchers suggest that individuals are willing to act against the greater collective good or to violate their own moral principles of fairness and justice if doing so will benefit a person for whom empathy is felt.

On a more positive note, aroused individuals in an empathetic manner may focus on the long-term welfare rather than just the short-term of those in need. Empathy-based socialization is very different from current practices directed toward inhibition of egoistic impulses through shaping, modeling and internalized guilt. Therapeutic programs built around facilitating altruistic impulses by encouraging perspective taking and empathetic feelings might enable individuals to develop more satisfactory interpersonal relations, especially in the long-term. At a societal level, experiments have indicated that empathy-induced altruism can be used to improve attitudes toward stigmatized groups, even used to improve racial attitudes, actions toward people with AIDS, the homeless and even convicts. Such resulting altruism has also been found to increase cooperation in competitive situations.

Types

Affective and Cognitive

Empathy is generally divided into two major components:

- **Affective empathy**, also called **emotional empathy**: the capacity to respond with an appropriate emotion to another's mental states. Our ability to empathize emotionally is based on emotional contagion: being affected by another's emotional or arousal state.
- **Cognitive empathy**: the capacity to understand another's perspective or mental state. The terms *cognitive empathy* and *theory of mind* are often used synonymously, but due to a lack of studies comparing theory of mind with types of empathy, it is unclear whether these are equivalent.

Although science has not yet agreed upon a precise definition of these constructs, there is consensus about this distinction.

Affective empathy can be subdivided into the following scales:

- **Empathic concern**: sympathy and compassion for others in response to their suffering.
- **Personal distress**: self-centered feelings of discomfort and anxiety in response to another's suffering. There is no consensus regarding whether personal distress is a basic form of empathy or instead does not constitute empathy. There may be a developmental aspect to this subdivision. Infants respond to the distress of others by getting distressed themselves; only when they are 2 years old do they start to respond in other-oriented ways, trying to help, comfort and share.

Cognitive empathy can be subdivided into the following scales:

- **Perspective taking**: the tendency to spontaneously adopt others' psychological perspectives.
- **Fantasy**: the tendency to identify with fictional characters.

Distribution

A difference in distribution between affective and cognitive empathy has been observed in various conditions. Psychopathy and narcissism have been associated with impairments in affective but not cognitive empathy, whereas bipolar disorder and borderline traits have been associated with deficits in cognitive but not affective empathy. Autism spectrum disorders have been associated with various combinations, including deficits in cognitive empathy as well as deficits in both cognitive and affective empathy. Schizophrenia, too, has been associated with deficits in both types of empathy. However, even in people without conditions such as these, the balance between affective and cognitive empathy varies.

A meta-analysis of recent fMRI studies of empathy confirmed that different brain areas are activated during affective–perceptual empathy and cognitive–evaluative empathy. Also, a study with patients with different types of brain damage confirmed the distinction between emotional and cognitive empathy. Specifically, the inferior frontal gyrus appears to be responsible for emotional empathy, and the ventromedial prefrontal gyrus seems to mediate cognitive empathy.

The Interpersonal Reactivity Index (IRI) is the only published measurement tool to date that accounts for a multi-dimensional assessment of empathy. It comprises a self-report questionnaire of 28 items, divided into four 7-item scales covering the above subdivisions of affective and cognitive empathy.

Somatic

- **Somatic empathy** is a physical reaction, probably based on mirror neuron responses, in the somatic nervous system.

Development

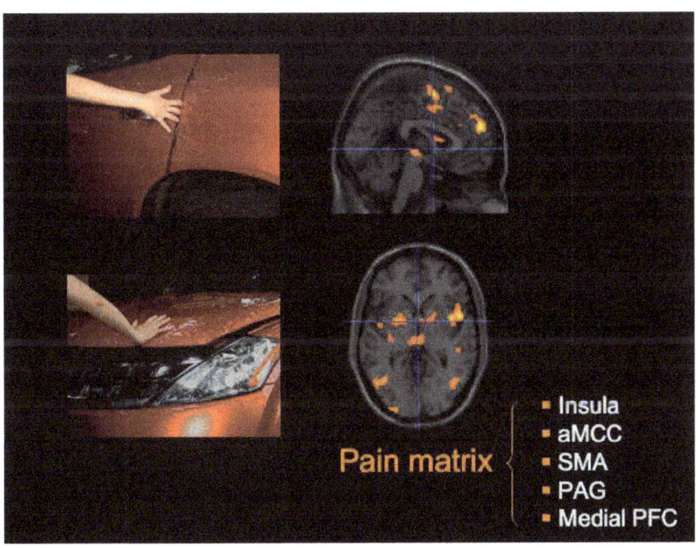

When children are shown videoclips with situations where they see people suffering pain by coincidence, neural circuits related to pain are being activated in their brain.

By the age of two years, children normally begin to display the fundamental behaviors of empathy by having an emotional response that corresponds with another person's emotional state. Even earlier, at one year of age, infants have some rudiments of empathy, in the sense that they understand that, just like their own actions, other people's actions have goals. Sometimes, toddlers will comfort others or show concern for them at as early an age as two. Also during the second year, toddlers will play games of falsehood or "pretend" in an effort to fool others, and this requires that the child know what others believe before he or she can manipulate those beliefs. In order to develop these traits, it is essential to expose your child to face-to-face interactions and opportunities and lead them away from a sedentary lifestyle.

According to researchers at the University of Chicago who used functional magnetic resonance imaging (fMRI), children between the ages of 7 and 12 years appear to be naturally inclined to feel empathy for others in pain. Their findingsare consistent with previous fMRI studies of pain empathy with adults. The research also found additional aspects of the brain were activated when youngsters saw another person intentionally hurt by another individual, including regions involved in moral reasoning.

Despite being able to show some signs of empathy, including attempting to comfort a crying baby, from as early as 18 months to two years, most children do not show a fully fledged theory of mind until around the age of four. Theory of mind involves the ability to understand that other people may have beliefs that are different from one's own, and is thought to involve the cognitive component of empathy. Children usually become capable of passing "false belief" tasks, considered to be a test for a theory of mind, around the age of four. Individuals with autism often find using a theory of mind very difficult (e.g. Baron-Cohen, Leslie & Frith, 1988; the Sally-Anne test).

Empathetic maturity is a cognitive structural theory developed at the Yale University School of Nursing and addresses how adults conceive or understand the personhood of patients. The theory, first applied to nurses and since applied to other professions, postulates three levels that have the properties of cognitive structures. The third and highest level is held to be a meta-ethical theory of the moral structure of care. Those adults operating with level-III understanding synthesize systems of justice and care-based ethics.

Individual Differences

Empathy in the broadest sense refers to a reaction of one individual to another's emotional state. Recent years have seen increased movement toward the idea that empathy occurs from motor neuron imitation. But, how do we account for individual differences in empathy? It cannot be said that empathy is a single unipolar construct but rather a set of constructs. In essence, not every individual responds equally and uniformly the same to various circumstances. The Empathic Concern scale assesses "other-oriented" feelings of sympathy and concern and the Personal Distress scale measures "self-oriented" feelings of personal anxiety and unease. The combination of these scales helps reveal those that might not be classified as empathetic and expands the narrow definition of empathy. Using this approach we can enlarge the basis of what it means to possess empathetic qualities and create a multi-faceted definition.

Behavioral and neuroimaging research show that two underlying facets of the personality dimensions Extraversion and Agreeableness (the Warmth-Altruistic personality profile) are associated with empathic accuracy and increased brain activity in two brain regions important for empathic processing (medial prefrontal cortex and temporoparietal junction).

Neurological Basis

Research in recent years has focused on possible brain processes underlying the experience of empathy. For instance, functional magnetic resonance imaging (fMRI) has been employed to investigate the functional anatomy of empathy. These studies have shown that observing another person's emotional state activates parts of the neuronal network involved in processing that same state in oneself, whether it is disgust, touch, or pain. The study of the neural underpinnings of empathy has received increased interest following the target paper published by Preston and Frans de Waal, following the discovery of mirror neurons in monkeys that fire both when the creature watches another perform an action as well as when they themselves perform it.

In their paper, they argue that attended perception of the object's state automatically activates neural representations, and that this activation automatically primes or generates the associated autonomic and somatic responses (idea of perception-action-coupling), unless inhibited. This mechanism is similar to the common coding theory between perception and action. Another recent study provides evidence of separate neural pathways activating reciprocal suppression in different regions of the brain associated with the performance of "social" and "mechanical" tasks. These findings suggest that the cognition associated with reasoning about the "state of another person's mind" and "causal/mechanical properties of inanimate objects" are neurally suppressed from occurring at the same time.

A recent meta-analysis of 40 fMRI studies found that affective empathy is correlated with increased activity in the insula while cognitive empathy is correlated with activity in the mid cingulate cortex and adjacent dorsomedial prefrontal cortex.

It has been suggested that mirroring-behavior in motor neurons during empathy may help duplicate feelings. Such sympathetic action may afford access to sympathetic feelings for another and, perhaps, trigger emotions of kindness, forgiveness.

Empathic Anger and Distress

Anger

Empathic anger is an emotion, a form of empathic distress. Empathic anger is felt in a situation where someone else is being hurt by another person or thing. It is possible to see this form of anger as a pro-social emotion.

Empathic anger has direct effects on both helping and punishing desires. Empathic anger can be divided into two sub-categories: trait empathic anger and state empathic anger.

The relationship between empathy and anger response towards another person has also been investigated, with two studies basically finding that the higher a person's perspective taking ability, the less angry they were in response to a provocation. Empathic concern did not, however, significantly predict anger response, and higher personal distress was associated with increased anger.

Distress

Empathic distress is feeling the perceived pain of another person. This feeling can be transformed into empathic anger, feelings of injustice, or guilt. These emotions can be perceived as pro-social, and some say they can be seen as motives for moral behavior.

Atypical response

Atypical empathic responses have been associated with autism and particular personality disorders such as psychopathy, borderline, narcissistic, and schizoid personality disorders; conduct disorder; schizophrenia; bipolar disorder; and depersonalization. Lack of empathy has also been associated with sex offenders. It was found that offenders that had been raised in an environment where they were shown a lack of empathy and had endured the same type of abuse, felt less empathy for their victims.

Autism

The interaction between empathy and autism is a complex and ongoing field of research. Several different factors are proposed to be at play.

A study of high-functioning adults with autism spectrum disorders found an increased prevalence of alexithymia, a personality construct characterized by the inability to recognize and articulate emotional arousal in oneself or others. Based on fMRI studies, alexithymia is responsible for a lack of empathy. The lack of empathic attunement inherent to alexithymic states may reduce quality and satisfaction of relationships. Recently, a study has shown that high-functioning adults with autism appear to have a range of responses to music similar to that of neurotypical individuals, including the deliberate use of music for mood management. Clinical treatment of alexithymia could involve using a simple associative learning process between musically induced emotions and their cognitive correlates. A study has suggested that the empathy deficits associated with the autism spectrum may be due to significant comorbidity between alexithymia and autism spectrum conditions rather than a result of social impairment.

One study found that, relative to typically developing children, high-functioning children with autism showed reduced mirror neuron activity in the brain's inferior frontal gyrus (pars opercularis) while imitating and observing emotional expressions. EEG evidence revealed that there was significantly greater mu suppression in the sensorimotor cortex of autistic individuals. Activity in this area was inversely related to symptom severity in the social domain, suggesting that a dysfunctional mirror neuron system may underlie social and communication deficits observed in autism, including impaired theory of mind and empathy. The mirror neuron system is essential for emotional empathy.

Previous studies have suggested that autistic individuals have an impaired theory of mind. Theory of mind is the ability to understand the perspectives of others.The terms cognitive empathy and theory of mind are often used synonymously, but due to a lack of studies comparing theory of mind with types of empathy, it is unclear whether these are equivalent. Theory of mind relies on structures of the temporal lobe and the pre-frontal cortex, and empathy, i.e. the ability to share the feelings of others, relies on the sensorimotor cortices as well as limbic and para-limbic structures. The lack of clear distinctions between theory of mind and empathy may have resulted in an incomplete understanding of the empathic abilities of those with Asperger syndrome; many reports on the empathic deficits of individuals with Asperger syndrome are actually based on impairments in theory of mind.

Studies have found that individuals on the autism spectrum self-report lower levels of empathic concern, show less or absent comforting responses toward someone who is suffering, and report equal or higher levels of personal distress compared to controls, which may be a result of high egocentrism found in autistic individuals. The combination in those on the autism spectrum of reduced empathic concern and increased personal distress may lead to the overall reduction of empathy. Professor Simon Baron-Cohen suggests that those with classic autism often lack both cognitive and affective empathy. Research also suggests that people with Asperger syndrome may have problems understanding others' perspectives in terms of theory of mind, but the average person with the condition demonstrates equal empathic concern as, and higher personal distress, than controls. The existence of individuals with heightened personal distress on the autism spectrum has been offered as an explanation as to why at least some people with autism would appear to have heightened emotional empathy,although increased personal distress may be an effect of heightened egocentrism, emotional empathy depends on mirror neuron activity (which, as described previously, has been found to be reduced in those with autism), and empathy in people on the autism spectrum is generally reduced. The empathy deficits present in autism spectrum disorders may be more indicative of impairments in the ability to take the perspective of others, while the empathy deficits in psychopathy may be more indicative of impairments in responsiveness to others' emotions.These "disorders of empathy" further highlight the importance of the ability to empathize by illustrating some of the consequences to disrupted empathy development.

The empathizing–systemizing theory (E-S) suggests that people may be classified on the basis of their capabilities along two independent dimensions, empathizing (E) and systemizing (S). These capabilities may be inferred through tests that measure someone's Empathy Quotient (EQ) and Systemizing Quotient (SQ). Five different "brain types" can be observed among the population based on the scores, which should correlate with differences at the neural level. In the E-S theory, autism and Asperger syndrome are associated with below-average empathy and average or above-average systemizing. The E-S theory has been extended into the Extreme Male Brain theory, which suggests that people with an autism spectrum condition are more likely to have an "Extreme Type S" brain type, corresponding with above-average systemizing but challenged empathy.

It has been shown that males are generally less empathetic than females. The Extreme Male Brain (EMB) theory proposes that individuals on the autistic spectrum are characterized by impairments in empathy due to sex differences in the brain: specifically, people with autism spectrum conditions show an exaggerated male profile. A study showed that some aspects of autistic neuroanatomy seem to be extremes of typical male neuroanatomy, which may be influenced by elevated levels of fetal testosterone rather than gender itself. Another study involving brain scans of 120 men and women suggested that autism affects male and female brains differently; females with autism had brains that appeared to be closer to those of non-autistic males than females, yet the same kind of difference was not observed in males with autism.

Psychopathy

Psychopathy is a personality disorder partly characterized by antisocial and aggressive behaviors, as well as emotional and interpersonal deficits including shallow emotions and a lack of remorse and empathy. The Diagnostic and Statistical Manual of Mental Disorders (DSM) and International Classification of Diseases (ICD) list antisocial personality disorder (ASPD) and dissocial personality disorder, stating that these have been referred to or include what is referred to as psychopathy.

A large body of research suggests that psychopathy is associated with atypical responses to distress cues (e.g. facial and vocal expressions of fear and sadness), including decreased activation of the fusiform and extrastriate cortical regions, which may partly account for impaired recognition of and reduced autonomic responsiveness to expressions of fear, and impairments of empathy. Studies on children with psychopathic tendencies have also shown such associations. The underlying biological surfaces for processing expressions of happiness are functionally intact in psychopaths, although less responsive than those of controls. The neuroimaging literature is unclear as to whether deficits are specific to particular emotions such as fear. Some recent fMRI studies have reported that emotion perception deficits in psychopathy are pervasive across emotions (positives and negatives).

A recent study on psychopaths found that, under certain circumstances, they could willfully empathize with others, and that their empathic reaction initiated the same way it does for controls. Psychopathic criminals were brain-scanned while watching videos of a person harming another individual. The psychopaths' empathic reaction initiated the same way it did for controls when they were instructed to empathize with the harmed individual, and the area of the brain relating to pain was activated when the psychopaths were asked to imagine how the harmed individual felt. The research suggests how psychopaths could switch empathy on at will, which would enable them to be both callous and charming. The team who conducted the study say it is still unknown how to transform this willful empathy into the spontaneous empathy most people have, though they propose it could be possible to bring psychopaths closer to rehabilitation by helping them to activate their "empathy switch". Others suggested that despite the results of the study, it remained unclear whether psychopaths' experience of empathy was the same as that of controls, and also questioned the possibility of devising therapeutic interventions that would make the empathic reactions more automatic.

Work conducted by Professor Jean Decety with large samples of incarcerated psychopaths offers additional insights. In one study, psychopaths were scanned while viewing video clips depicting people being intentionally hurt. They were also tested on their responses to seeing short videos of facial expressions of pain. The participants in the high-psychopathy group exhibited significantly less activation in the ventromedial prefrontal cortex, amygdala and periaqueductal gray parts of the brain, but more activity in the striatum and the insula when compared to control participants. In a second study, individuals with psychopathy exhibited a strong response in pain-affective brain regions when taking an imagine-self perspective, but failed to recruit the neural circuits that were activated in controls during an imagine-other perspective—in particular the ventromedial prefrontal cortex and amygdala—which may contribute to their lack of empathic concern.

It was predicted that people who have high levels of psychopathy would have sufficient levels of cognitive empathy but would lack in their ability to use affective empathy. People that scored highly on psychopathy measures were less likely to portray affective empathy. There was a strong negative correlation showing that psychopathy and affective empathy are correspond strongly. The DANVA-2 portrayed those who scored highly on the psychopathy scale do not lack in recognising emotion in facial expressions. Therefore, individuals who have high scores on psychopathy and do not lack in perspective-talking ability but do lack in compassion and the negative incidents that happen to others.

Despite studies suggesting deficits in emotion perception and imagining others in pain, professor Simon Baron-Cohen claims psychopathy is associated with intact cognitive empathy, which would imply an intact ability to read and respond to behaviors, social cues and what others are feeling. Psychopathy is, however, associated with impairment in the other major component of empathy—affective (emotional) empathy—which includes the ability to *feel* the suffering and emotions of others (what scientists would term as emotional contagion), and those with the condition are therefore not distressed by the suffering of their victims. Those with autism, on the other hand, are often impaired in both affective and cognitive empathy.

Other Conditions

Research indicates atypical empathic responses are also correlated with a variety of other conditions.

Borderline personality disorder is characterized by extensive behavioral and interpersonal difficulties that arise from emotional and cognitive dysfunction.Dysfunctional social and interpersonal behavior has been shown to play a crucial role in the emotionally intense way people with borderline personality disorder react. While individuals with borderline personality disorder may show their emotions too much, several authors have suggested that they might have a compromised ability to reflect upon mental states (impaired cognitive empathy), as well as an impaired theory of mind. People with borderline personality disorder have been shown to be very good at recognizing emotions in people's faces, suggesting increased empathic capacities. It is, therefore, possible that impaired cognitive empathy (the capacity for understanding another person's experience and perspective) may account for borderline personality disorder individuals' tendency for interpersonal dysfunction, while "hyper-emotional empathy" may account for the emotional over-reactivity observed in these individuals. One primary study confirmed that patients with borderline personality disorder were significantly impaired in cognitive empathy, yet there was no sign of impairment in affective empathy.

One diagnostic criterion of narcissistic personality disorder is a lack of empathy and an unwillingness or inability to recognize or identify with the feelings and needs of others.

Characteristics of schizoid personality disorder include emotional coldness, detachment, and impaired affect corresponding with an inability to be empathetic and sensitive towards others.

A study conducted by Jean Decety and colleagues at the University of Chicago demonstrated that subjects with aggressive conduct disorder elicit atypical empathic responses to viewing others in pain. Subjects with conduct disorder were at least as responsive as controls to the pain of others but, unlike controls, subjects with conduct disorder showed strong and specific activation of the amygdala and ventral striatum (areas that enable a general arousing effect of reward), yet impaired activation of the neural regions involved in self-regulation and metacognition (including moral reasoning), in addition to diminished processing between the amygdala and the prefrontal cortex.

Schizophrenia is characterized by impaired affective empathy, as well as severe cognitive and empathy impairments as measured by the Empathy Quotient (EQ). These empathy impairments are also associated with impairments in social cognitive tasks.

Bipolar individuals have been observed to have impaired cognitive empathy and theory of mind, but increased affective empathy. Despite cognitive flexibility being impaired, planning behavior is intact. It has been suggested that dysfunctions in the prefrontal cortex could result in the impaired cognitive empathy, since impaired cognitive empathy has been related with neurocognitive task performance involving cognitive flexibility.

Lieutenant Colonel Dave Grossman, in his book *On Killing*, suggests that military training artificially creates depersonalization in soldiers, suppressing empathy and making it easier for them to kill other human beings.

Practical Issues

The capacity to empathize is a revered trait in society. Empathy is considered a motivating factor for unselfish, prosocial behavior, whereas a lack of empathy is related to antisocial behavior.

Proper empathic engagement helps an individual understand and anticipate the behavior of another. Apart from the automatic tendency to recognize the emotions of others, one may also deliberately engage in empathic reasoning. Two general methods have been identified here. An individual may simulate fictitious versions of the beliefs, desires, character traits and context of another individual to see what emotional feelings it provokes. Or, an individual may simulate an emotional feeling and then access the environment for a suitable reason for the emotional feeling to be appropriate for that specific environment.

Some research suggests that people are more able and willing to empathize with those most similar to themselves. In particular, empathy increases with similarities in culture and living conditions. Empathy is more likely to occur between individuals whose interaction is more frequent. (See Levenson and Reuf 1997 and Hoffman 2000: 62). A measure of how well a person can infer the specific content of another person's thoughts and feelings has been developed by William Ickes (1997, 2003). Ickes and his colleagues have developed a video-based method to measure empathic accuracy and have used this method to study the empathic inaccuracy of maritally aggressive and abusive spouses, among other topics.

There are concerns that the empathizer's own emotional background may affect or distort what emotions they perceive in others (e.g. Goleman 1996: p. 104). Empathy is not a process that is likely to deliver certain judgments about the emotional states of others. It is a skill that is gradually developed throughout life, and which improves the more contact we have with the person with whom one empathizes. Accordingly, any knowledge gained of the emotions of the other must be revisable in light of further information. Empathizers report finding it easier to take the perspective of another person when they have experienced a similar situation, as well as experience greater empathic emotion. Research regarding whether similar past experience makes the empathizer more accurate is mixed.

Ethical Issues

The extent to which a person's emotions are publicly observable, or mutually recognized as such has significant social consequences. Empathic recognition may or may not be welcomed or socially desirable. This is particularly the case where we recognize the emotions that someone has towards us during real time interactions. Based on a metaphorical affinity with touch, philosopher Edith Wyschogrod claims that the proximity entailed by empathy increases the potential vulnerability of either party. The appropriate role of empathy in our dealings with others is highly dependent on the circumstances. For instance, Tania Singer claims that clinicians or caregivers must take care not to be too sensitive to the emotions of others, to over-invest their own emotions, at the risk of draining away their own resourcefulness. Furthermore, an awareness of the limitations of empathic accuracy is prudent in a caregiving situation.

Disciplinary Approaches

Philosophy

Ethics

In his 2008 book, *How to Make Good Decisions and Be Right All the Time:Solving the Riddle of Right and Wrong*, writer Iain King presents two reasons why empathy is the "essence" or "DNA" of right and wrong. First, he argues that empathy uniquely has all the characteristics we can know about an ethical viewpoint– including that it is "partly self-standing", and so provides a source of motivation that is partly within us and partly outside, as moral motivations seem to be. This allows empathy-based judgements to have sufficient distance from a personal opinion to count as "moral". His second argument is more practical: he argues, "Empathy for others really is the route to value in life", and so the means by which a selfish attitude can become a moral one. By using empathy as the basis for a system of ethics, King is able to reconcile ethics based on consequences with virtue-ethics and act-based accounts of right and wrong. His empathy-based system has been taken up by some Buddhists, and is used to address some practical problems, such as when to tell lies, and how to develop culturally-neutral rules for romance.

In the 2007 book *The Ethics of Care and Empathy*, philosopher Michael Slote introduces a theory of care-based ethics that is grounded in empathy. His claim is that moral motivation does, and should, stem from a basis of empathic response. He claims that our natural reaction to situations of moral significance are explained by empathy. He explains that the limits and obligations of empathy and in turn morality are natural. These natural obligations include a greater empathic, and moral obligation to family and friends, along with an account of temporal and physical distance. In situations of close temporal and physical distance, and with family or friends, our moral obligation seems stronger to us than with strangers at a distance naturally. Slote explains that this is due to empathy and our natural empathic ties. He further adds that actions are wrong if and only if they reflect or exhibit a deficiency of fully developed empathic concern for others on the part of the agent.

Phenomenology

In phenomenology, empathy describes the experience of something from the other's viewpoint, without confusion between self and other. This draws on the sense of agency. In the most basic sense, this is the experience of the other's body and, in this sense, it is an experience of "my body over there". In most other respects, however, the experience is modified so that what is experienced is experienced as being the other's experience; in experiencing empathy, what is experienced is not "my" experience, even though *I* experience it. Empathy is also considered to be the condition of intersubjectivity and, as such, the source of the constitution of objectivity.

Some postmodern historians such as Keith Jenkins in recent years have debated whether or not it is possible to empathize with people from the past. Jenkins argues that empathy only enjoys such a privileged position in the present because it corresponds harmoniously with the dominant liberal discourse of modern society and can be connected to John Stuart Mill's concept of reciprocal freedom. Jenkins argues the past is a foreign country and as we do not have access to the epistemological conditions of by gone ages we are unable to empathize.

It is impossible to forecast the effect of empathy on the future. A past subject may take part in the present by the so-called historic present. If we watch from a fictitious past, can tell the present with the future tense, as it happens with the trick of the false prophecy. There is no way of telling the present with the means of the past.

Evolution

An increasing number of studies in animal behavior and neuroscience claim that empathy is not restricted to humans, and is in fact as old as the mammals, or perhaps older. Examples include dolphins saving humans from drowning or from shark attacks. Professor Tom White suggests that reports of cetaceans having three times as many spindle cells — the nerve cells that convey empathy — in their brains as we do might mean these highly-social animals have a great awareness of one another's feelings.

A multitude of behaviors has been observed in primates, both in captivity and in the wild, and in particular in bonobos, which are reported as the most empathetic of all the primates. A recent study has demonstrated prosocial behavior elicited by empathy in rodents.

Rodents have been shown to demonstrate empathy for cagemates (but not strangers) in pain. One of the most widely read studies on the evolution of empathy, which discusses a neural perception-action mechanism (PAM), is the one by Stephanie Preston and de Waal. This review postulates a bottom-up model of empathy that ties together all levels, from state matching to perspective-taking. For University of Chicago neurobiologist Jean Decety, [empathy] is not specific to humans. He argues that there is strong evidence that empathy has deep evolutionary, biochemical, and neurological underpinnings, and that even the most advanced forms of empathy in humans are built on more basic forms and remain connected to core mechanisms associated with affective communication, social attachment, and parental care. Core neural circuits that are involved in empathy and caring include the brainstem, the amygdala, hypothalamus, basal ganglia, insula and orbitofrontal cortex.

Context Evolution Problems

Since all definitions of empathy involves an element of for others, all distinctions between egoism and empathy fail at least for beings lacking self-awareness. Since the first mammals lacked a self-aware distinction between self and other, as shown by most mammals failing at mirror tests, the first mammals or anything more evolutionarily primitive than them cannot have had a context of default egoism requiring an empathy mechanism to be transcended. However, there are numerous examples in artificial intelligence research showing that simple reactions can carry out de facto functions the agents have no concept of, so this does not contradict evolutionary explanations of parental care. However, such mechanisms would be unadapted to self-other distinction and beings already dependent on some form of behavior benefitting each other or their offspring would never be able to evolve a form of self-other distinction that necessitated evolution of specialized non-preevolved and non-preevolvable mechanisms for retaining empathic behavior in the presence of self-other distinction, and so a fundamental neurological distinction between egoism and empathy cannot exist in any species.

Psychotherapy

Heinz Kohut is the main introducer of the principle of empathy in psychoanalysis. His principle applies to the method of gathering unconscious material. The possibility of not applying the principle is granted in the cure, for instance when you must reckon with another principle, that of reality. Developing skills of empathy is often a central theme in the recovery process for drug addicts.

In evolutionary psychology, attempts at explaining pro-social behavior often mention the presence of empathy in the individual as a possible variable. While exact motives behind complex social behaviors are difficult to distinguish, the "ability to put oneself in the shoes of another person and experience events and emotions the way that person experienced them" is the definitive factor for truly altruistic behavior according to Batson's empathy-altruism hypothesis. If empathy is not felt, social exchange (what's in it for me?) supersedes pure altruism, but if empathy is felt, an individual will help by actions or by word, regardless of whether it is in their self-interest to do so and even if the costs outweigh potential rewards.

Education

An important target of the method Learning by teaching (LbT) is to train systematically and, in each lesson, teach empathy. Students have to transmit new content to their classmates, so they have to reflect continuously on the mental processes of the other students in the classroom. This way it is possible to develop step-by-step the students' feeling for group reactions and networking. Carl R. Rogers pioneered research in effective psychotherapy and teaching which espoused that empathy coupled with unconditional positive regard or caring for students and authenticity or congruence were the most important traits for a therapist or teacher to have. Other research and publications by Tausch, Aspy, Roebuck. Lyon, and meta-analyses by Cornelius-White, corroborated the importance of these person-centered traits.

Business and management

In the 2009 book *Wired to Care*, strategy consultant Dev Patnaik argues that a major flaw in contemporary business practice is a lack of empathy inside large corporations. He states that lacking any sense of empathy, people inside companies struggle to make intuitive decisions and often get fooled into believing they understand their business if they have quantitative research to rely upon. Patnaik claims that the real opportunity for companies doing business in the 21st Century is to create a widely held sense of empathy for customers, pointing to Nike, Harley-Davidson, and IBM as examples of "Open Empathy Organizations". Such institutions, he claims, see new opportunities more quickly than competitors, adapt to change more easily, and create workplaces that offer employees a greater sense of mission in their jobs. In the 2011 book *The Empathy Factor*, organizational consultant Marie Miyashiro similarly argues the value of bringing empathy to the workplace, and offers Nonviolent Communication as an effective mechanism for achieving this. In studies by the Management Research Group, empathy was found to be the strongest predictor of ethical leadership behavior out of 22 competencies in its management model, and empathy was one of the three strongest predictors of senior executive effectiveness.

Intercultural

Intercultural empathy is the ability to perceive the world as it is perceived by a culture different from the subject's own. Empathy interculturally regards a variety of issues, such as the approach to time perception (deadlines, temporal precision, perspectives on time), how to negotiate with people from different cultures and organizations, and be able to integrate different communication styles and cultures. A specific model in the literature identifies four levels of empathy. The model, developed by the European researcher

Daniele Trevisani, specifies the need for at least four types of understanding for empathy in intercultural settings:

- Behavioral empathy: the ability to understand why the behavior is adopted and the chains of related behaviors.
- Emotional empathy: being able to feel the emotions experienced by others, even in cultures different from one's own, how emotions are associated to people, objects, events, situations, in different cultures.
- Relational empathy: understanding the map of the relations of the subject mapping with whom the subject relates whether voluntarily or compulsorily, who has to deal with that subject in order to decide, in work or life, what is his map of "significant others ", who can affects his/her professional and life decisions.
- Cognitive empathy (understanding of different cognitive or prototypes): understanding the cognitive prototypes active in a person, the culturally different beliefs that generate the observable behaviors, the mental structures that the individuals own ("memetic background").

As US researchers William Weeks, Paul Pedersen et al. state, developing intercultural empathy enables the interpretation of experiences or perspectives from more than one worldview. Intercultural empathy can also improve self-awareness and critical awareness of one's own interaction style as conditioned by one's cultural views and promote a view of self-as-process.

Measurement

Research into the measurement of empathy has sought to answer a number of questions: who should be carrying out the measurement? What should pass for empathy and what should be discounted? What unit of measure (UOM) should be adopted and to what degree should each occurrence precisely match that UOM are also key questions that researchers have sought to investigate.

Researchers have approached the measurement of empathy from a number of perspectives.

Behavioral measures normally involve raters assessing the presence or absence of certain either predetermined or ad-hoc behaviors in the subjects they are monitoring. Both verbal and non-verbal behaviors have been captured on video by experimenters such as Truax (1967b). Other experimenters, including Mehrabian and Epstein (1972), have required subjects to comment upon their own feelings and behaviors, or those of other people involved in the experiment, as indirect ways of signaling their level of empathic functioning to the raters.

Physiological responses tend to be captured by elaborate electronic equipment that has been physically connected to the subject's body. Researchers then draw inferences about that person's empathic reactions from the electronic readings produced (e.g. Levenson and Ruef, 1992; Leslie et al., 2004).

Bodily or "somatic" measures can be looked upon as behavioral measures at a micro level. Their focus is upon measuring empathy through facial and other non-verbally expressed reactions in the empathizer. These changes are presumably underpinned by physiological changes brought about by some form of "emotional contagion" or mirroring (e.g. Levenson and Ruef, 1992*; Leslie et al., 2004*). It should be pointed out that these reactions, whilst appearing to reflect the internal emotional state of the empathizer, could also, if the stimulus incident lasted more than the briefest period, be reflecting the results of emotional reactions that are based upon more pieces of thinking through (cognitions) associated with role-taking ("if I were him I would feel ...").

Paper-based indices involve one or more of a variety of methods of responding. In some experiments, subjects are required to watch video scenarios (either staged or authentic) and to make written responses which are then assessed for their levels of empathy (e.g. Geher, Warner and Brown, 2001); scenarios are sometimes also depicted in printed form (e.g. Mehrabian and Epstein, 1972). Measures also frequently require subjects to self-report upon their own ability or capacity for empathy, using Likert-style numerical responses to a printed questionnaire that may have been designed to tap into the affective, cognitive-affective or largely cognitive substrates of empathic functioning. Some questionnaires claim to have been able to tap into both cognitive and affective substrates (e.g. Davis, 1980). More recent paper-based tools include The Empathy Quotient (EQ) created by Baron-Cohen and Wheelwright which comprises a self-report questionnaire consisting of 60 items.

For the very young, picture or puppet-story indices for empathy have been adopted to enable even very young, pre-school subjects to respond without needing to read questions and write answers (e.g. Denham and Couchoud, 1990). Dependent variables (variables that are monitored for any change by the experimenter) for younger subjects have included self reporting on a 7-point smiley face scale and filmed facial reactions (Barnett, 1984).

A certain amount of confusion exists about how to measure empathy. These may be rooted in another problem: deciding what empathy is and what it is not. In general, researchers have until now been keen to pin down a singular definition of empathy which would allow them to design a measure to assess its presence in an exchange, in someone's repertoire of behaviors or within them as a latent trait. As a result, they have been frequently forced to ignore the richness of the empathic process in favor of capturing surface, explicit self-report or third-party data about whether empathy between two people was present or not. In most cases, instruments have unfortunately only yielded information on whether someone had the potential to demonstrate empathy (Geher et al., 2001)*. Gladstein (1987)summarizes the position noting that empathy has been measured from the point of view of the empathizer, the recipient for empathy and the third-party observer. He suggests that since the multiple measures used have produced results that bear little relation to one another, researchers should refrain from making comparisons between scales that are in fact measuring different things. He suggests that researchers should instead stipulate what kind of empathy they are setting out to measure rather than simplistically stating that they are setting out to measure the unitary phenomenon "empathy"; a view more recently endorsed by Duan and Hill (1996).

In the field of medicine, a measurement tool for carers is the *Jefferson Scale of Physician Empathy, Health Professional Version (JSPE-HP)*. At least one study using this tool with health sciences' students has found that levels of empathy are greater amongst females than males, and also are greater amongst older students than younger students.

The Interpersonal Reactivity Index (IRI) is the only published measurement tool accounting for a multi-dimensional assessment of empathy, consisting of a self-report questionnaire of 28 items, divided into four 7-item scales covering the subdivisions of affective and cognitive empathy.

Other Animals

Research has shown that the ability of empathy in other species is attainable. Many instances of empathy have been recorded throughout many species, including (but not limited to) canines, felines, dolphins, primates, rats and mice. In animals, empathy-related responding could have an ulterior motive such as survival, the sharing of food, companionship and pack-oriented mentality. It is certainly difficult to understand an animal's intention behind an empathic response. Many researchers maintain that applying the term empathy in general to animal behavior is an act of anthropomorphism.

Researchers Zanna Clay and Frans de Waal studied the socio-emotional development of the bonobo chimpanzee. They focused on the interplay of numerous skills such as empathy-related responding, and how different rearing backgrounds of the juvenile bonobo affected their response to stressful events, related to themselves (loss of a fight) and of stressful events of others. It was found that the bonobos sought out body contact as a coping mechanism with one another. A finding of this study was that the bonobos sought out more body contact after watching a distressing event upon the other bonobos rather than their individually experienced stressful event. Mother-reared bonobos as opposed to orphaned bonobos sought out more physical contact after a stressful event happened to another. This finding shows the importance of mother-child attachment and bonding, and how it may be crucial to successful socio-emotional development, such as empathic-like behaviors.

Empathic-like responding has been observed in chimpanzees in various different aspects of their natural behaviors. For example, chimpanzees are known to spontaneously contribute comforting behaviors to victims of aggressive behavior in natural and unnatural settings, a behavior recognized as consolation. Researchers Teresa Romero and co-workers observed these empathic and sympathetic-like behaviors in chimpanzees at two separate outdoor housed groups.The act of consolation was observed in both of the groups of chimpanzees. This behavior is found in humans, and particularly in human infants. Another similarity found between chimpanzees and humans is that empathic-like responding was disproportionately provided to individuals of kin. Although comforting towards non-family chimpanzees was also observed, as with humans, chimpanzees showed the majority of comfort and concern to close/loved ones. Another similarity between chimpanzee and human expression of empathy is that females provided more comfort than males on average. The only exception to this discovery was that high-ranking males showed as much empathy-like behavior as their female counterparts. This is believed to be because of policing-like behavior and the authoritative status of high-ranking male chimpanzees.

It is thought that species that possess a more intricate and developed prefrontal cortex have more of an ability of experiencing empathy. It has however been found that empathic and altruistic responses may also be found in sand dwelling Mediterranean ants. Researcher Hollis studied the *Cataglyphis cursor* sand dwelling Mediterranean ant and their rescue behaviors by ensnaring ants from a nest in nylon threads and partially buried beneath the sand. The ants not ensnared in the nylon thread proceeded to attempt to rescue their nest mates by sand digging, limb pulling, transporting sand away from the trapped ant, and when efforts remained unfruitful, began to attack the nylon thread itself; biting and pulling apart the threads. Similar rescue behavior was found in other sand-dwelling Mediterranean ants, but only *Cataglyphis floricola* and *Lasius grandis* species of ants showed the same rescue behaviors of transporting sand away from the trapped victim and directing attention towards the nylon thread. It was observed in all ant species that rescue behavior was only directed towards nest mates. Ants of the same species from different nests were treated with aggression and were continually attacked and pursued, which speaks to the depths of ants discriminative abilities. This study brings up the possibility that if ants have the capacity for empathy and/or altruism, these complex processes may be derived from primitive and simpler mechanisms.

Canines have been hypothesized to share empathic-like responding towards human species. Researchers Custance and Mayer put individual dogs in an enclosure with their owner and a stranger. When the participants were talking or humming, the dog showed no behavioral changes, however when the participants were pretending to cry, the dogs oriented their behavior toward the person in distress whether it be the owner or stranger. The dogs approached the participants when crying in a submissive fashion, by sniffing, licking and nuzzling the distressed person. The dogs did not approach the participants in the usual form of excitement, tail wagging or panting. Since the dogs did not direct their empathic-like responses only towards their owner, it is hypothesized that dogs generally seek out humans showing distressing body behavior. Although this could insinuate that dogs have the cognitive capacity for empathy, this could also mean that domesticated dogs have learned to comfort distressed humans through generations of being rewarded for that specific behavior.

When witnessing chicks in distress, domesticated hens, *Gallus gallus domesticus* show emotional and physiological responding. Researchers Edgar, Paul and Nicol found that in conditions where the chick was susceptible to danger, the mother hens heart rate increased, vocal alarms were sounded, personal preening decreased and body temperature increased. This responding happened whether or not the chick felt as if they were in danger. Mother hens experienced stress-induced hyperthermia only when the chick's behavior correlated with the perceived threat. Animal maternal behavior may be perceived as empathy, however, it could be guided by the evolutionary principles of survival and not emotionality.

REFERENCES

1. Panksepp, J. (1982). "Toward a general psychobiological theory of emotions". Behavioral and Brain Sciences. 5 (3): 407–422. doi:10.1017/S0140525X00012759.
2. "Emotions help animals to make choices (press release)". University of Bristol. 2010. Retrieved October 26, 2013.
3. Jacky Turner; Joyce D'Silva, eds. (2006). Animals, Ethics and Trade: The Challenge of Animal Sentience. Earthscan. ISBN 9781844072545. Retrieved October 26, 2013.

4. Wong, K. (2013). "How to identify grief in animals". Scientific American. Retrieved October 26, 2013.

5. Rygula, R; Pluta, H; P, Popik (2012). "laughing rats are optimistic". PLoS ONE. 7 (12): e51959. doi:10.1371/journal.pone.0051959. PMC 3530570Freely accessible. PMID 23300582.

6. Douglas, C.; Bateson, M.last2=Bateson; Walsh, C.; Béduéc, A.; Edwards, S.A. (2012). "Environmental enrichment induces optimistic cognitive biases in pigs". Applied Animal Behaviour Science. 139 (1–2): 65–73. doi:10.1016/j.applanim.2012.02.018.

7. Bateson, M.; Desire, S.; Gartside, S.E.; Wright, G.A. (2011). "Agitated honeybees exhibit pessimistic cognitive biases". Current Biology. 21 (12): 1070–1073. doi:10.1016/j.cub.2011.05.017. PMID 21636277.

8. Merriam-Webster (2004). The Merriam-Webster dictionary (11th ed.). Springfield, MA: Author.

9. Fox, E. (2008). Emotion Science: An Integration of Cognitive and Neuroscientific Approaches. Palgrave MacMillan. pp. 16–17. ISBN 978-0-230-00517-4.

10. Ekman, P. (1992). "An argument for basic emotions". Cognition and Emotion. 6 (3): 169–200. doi:10.1080/02699939208411068.

11. Barnard, S., Matthews, L., Messori, S., Podaliri-Vulpiani, M. and Ferri, N. (2015). "Laterality as an indicator of emotional stress in ewes and lambs during a separation test". Animal Cognition. 19: 1–8. doi:10.1007/s10071-015-0928-3.

12. Handel, S. "Classification of Emotions". Retrieved 30 April 2012.

13. Dawkins, M. (2000). "Animal minds and animal emotions". American Zoologist. 40 (6): 883–888. doi:10.1668/0003-1569(2000)040[0883:amaae]2.0.co;2.

14. Dixon, B. (2001). "Animal emotions. Ethics and the Environment". Ethics. 6 (2): 22–30. doi:10.2979/ete.2001.6.2.22.

15. Moussaieff Masson, J.; McCarthy, S. (1996). When Elephants Weep: The Emotional Lives of Animals. Delta. ISBN 0-385-31428-0.

16. D.S. Mills; J.N. Marchant-Forde, eds. (2010). The Encyclopedia of Applied Animal Behaviour and Welfare. CABI. ISBN 0851997244.

17. Morgan, C.L. (1903). An Introduction to Comparative Psychology (2nd ed.). W. Scott, London. p. 59.

18. Darwin, C. (1872). The Expression Of The Emotions In Man And Animals. University of Chicago Press, Chicago.

19. Hess, U. and Thibault, P., (2009). Darwin and emotion expression. American Psychological Association, 64(2): 120-128.

20. Scruton, R; Tyler, A. (2001). "Debate: Do animals have rights?". The Ecologist. 31 (2): 20–23.

21. Oately, K.; Jenkins, J.M. (1996). Understanding Emotions. Blackwell Publishers. Malden, MA. ISBN 1-55786-495-0.

22. Paul, E; Harding, E; Mendl, M (2005). "Measuring emotional processes in

animals: the utility of a cognitive approach". Neuroscience and Biobehavioral Reviews. 29 (3): 469–491. doi:10.1016/j.neubiorev.2005.01.002. PMID 15820551.

23. Haselton, M. G.; Nettle, D. & Andrews, P. W. (2005). The evolution of cognitive bias. In D. M. Buss (Ed.), The Handbook of Evolutionary Psychology: Hoboken, NJ, US: John Wiley & Sons Inc. pp. 724–746.
24. Bless, H.; Fiedler, K. & Strack, F. (2004). Social cognition: How individuals construct social reality. Hove and New York: Psychology Press. p. 2.
25. Harding, EJ; Paul, ES; Mendl, M (2004). "Animal behaviour: cognitive bias and affective state". Nature. 427 (6972): 312. doi:10.1038/427312a. PMID 14737158.
26. Sherwin, C.M.; Olsson, I.A.S. (2004). "Housing conditions affect self-administration of anxiolytic by laboratory mice.". Animal Welfare. 13: 33–38.
27. Fajardo, C.; et al. (4 March 2008). "Von Economo neurons are present in the dorsolateral (dysgranular) prefrontal cortex of humans". Neuroscience Letters. 435 (3): 215–218. doi:10.1016/j.neulet.2008.02.048. PMID 18355958.
28. Hof, P.R.; Van Der Gucht, E. (2007). "Structure of the cerebral cortex of the humpback whale, Megaptera novaeangliae (Cetacea, Mysticeti, Balaenopteridae)". Anatomical Record, Part A. 290 (1): 1–31. doi:10.1002/ar.20407. PMID 17441195.
29. Coghlan, A. (27 November 2006). "Whales boast the brain cells that 'make us human'". New Scientist.
30. Butti, C; Sherwood, CC; Hakeem, AY; Allman, JM; Hof, PR (July 2009). "Total number and volume of Von Economo neurons in the cerebral cortex of cetaceans.". The Journal of Comparative Neurology. 515 (2): 243–59. doi:10.1002/cne.22055. PMID 19412956.
31. Hakeem, A. Y.; Sherwood, C. C.; Bonar, C. J.; Butti, C.; Hof, P. R.; Allman, J. M. (2009). "Von Economo neurons in the elephant brain". The Anatomical Record (Hoboken). 292 (2): 242–8. doi:10.1002/ar.20829. PMID 19089889.
32. Panksepp, J; Burgdorf, J (2003). ""Laughing" rats and the evolutionary antecedents of human joy?". Physiology and Behavior. 79 (3): 533–547. doi:10.1016/s0031-9384(03)00159-8. PMID 12954448.
33. Knutson, B; Burgdorf, J; Panksepp, J (2002). "Ultrasonic vocalizations as indices of affective states in rats". Psychological Bulletin. 128 (6): 961–977. doi:10.1037/0033-2909.128.6.961. PMID 12405139.
34. Panksepp, J; Burgdorf, J (2000). "50-kHz chirping (laughter?) in response to conditioned and unconditioned tickle-induced reward in rats: effects of social housing and genetic variables". Behavioural Brain Research. 115 (1): 25–38. doi:10.1016/s0166-4328(00)00238-2. PMID 10996405.
35. Boissy, A.; et al. (2007). "Assessment of positive emotions in animals to improve their welfare". Physiology & Behavior. 92 (3): 375–397. doi:10.1016/j.physbeh.2007.02.003.
36. Elfenbein, H.A.; Ambady, N. (2002). "On the universality and cultural specificity of emotion recognition: A meta-analysis". Psychological Bulletin. 128 (2): 203–235. doi:10.1037/0033-2909.128.2.203. PMID 11931516.

37. Winford, J.N. (2007). "Almost human, and sometimes smarter". New York Times. Retrieved October 26, 2013.

38. "Mama gorilla won't let go of her dead baby". Associated Press. 2008. Retrieved October 26, 2013.

39. McGraw, C. (1985). "Gorilla's Pets: Koko Mourns Kittens Death". Los Angeles Times. Retrieved October 26, 2013.

40. Wechkin, S.; Masserman, J.H.; Terris, W. (1964). "Shock to a conspecific as an aversive stimulus". Psychonomic Science. 1: 47–48. doi:10.3758/bf03342783.

41. Masserman, J.; Wechkin, M.S.; Terris, W. (1964). "Altruistic behavior in rhesus monkeys". American Journal of Psychiatry. 121 (6): 584–585. doi:10.1176/ajp.121.6.584. PMID 14239459.

42. De Waal, F.B.M. & Aureli, F. (1996). Consolation, reconciliation, and a possible cognitive difference between macaques and chimpanzees. In A.E. Russon, K.A. Bard, and S.T. Parker (Eds.), Reaching Into Thought: The Minds Of The Great Apes (pp. 80-110). Cambridge: Cambridge University Press.

43. Watts, D.P., Colmenares, F., & Arnold, K. (2000). Redirection, consolation, and male policing: How targets of aggression interact with bystanders. In F. Aureli and F.B.M. de Waal (Eds.), Natural Conflict Resolution (pp. 281-301). Berkeley: University of California Press.

44. Parr, L.A. (2001). "Cognitive and physiological markers of emotional awareness in chimpanzees". Animal Cognition. 4 (3–4): 223–229. doi:10.1007/s100710100085. PMID 24777512.

45. Panksepp, J. (1998). Affective. Neuroscience: The Foundation of Human and Animal Emotions. Oxford University Press, New York. p. 480.

46. Panksepp, J.B.; Lahvis, G.P. (2011). "Rodent empathy and affective neuroscience". Neuroscience and Biobehavioral Reviews. 35 (9): 1864–1875. doi:10.1016/j.neubiorev.2011.05.013. PMID 21672550.

47. Rice, G.E.; Gainer, P (1962). "Altruism in the albino rat". Journal of Comparative and Physiological Psychology. 55: 123–125. doi:10.1037/h0042276. PMID 14491896.

48. Mirsky, I.A.; Miller, R.E.; Murphy, J.B. (1958). "The communication of affect in rhesus monkeys I. An experimental method". Journal of the American Psychoanalytic Association. 6 (3): 433–441. doi:10.1177/000306515800600303. PMID 13575267.

49. Langford, D.J.; Crager, S.E.; Shehzad, Z.; Smith, S.B.; Sotocinal, S.G.; Levenstadt, J.S.; Mogil, J.S. (2006). "Social modulation of pain as evidence for empathy in mice". Science. 312 (5782): 1967–1970. doi:10.1126/science.1128322. PMID 16809545.

50. Reimerta, I.; Bolhuis, J.E; Kemp, B.; Rodenburg., T.B. (2013). "Indicators of positive and negative emotions and emotional contagion in pigs". Physiology and Behavior. 109: 42–50. doi:10.1016/j.physbeh.2012.11.002. PMID 23159725.

51. Atsak, P.; Ore, M; Bakker, P.; Cerliani, L.; Roozendaal, B.; Gazzola, V.; Moita,

M.; Keysers, C. (2011). "Experience modulates vicarious freezing in rats: a model for empathy". PLoS ONE. 6 (7): e21855. doi:10.1371/journal.pone.0021855. PMC 3135600Freely accessible. PMID 21765921.

52. Kavaliers, M.; Colwell, D.D.; Choleris, E. (2003). "Learning to fear and cope with a natural stressor: individually and socially acquired corticosterone and avoidance responses to biting flies". Hormonal Behavior. 43: 99–107. doi:10.1016/s0018-506x(02)00021-1.

53. Kim, E.J.; Kim, E.S.; Covey, E.; Kim, J.J. (2010). "Social transmission of fear in rats: the role of 22-kHz ultrasonic distress vocalization". PLoS ONE. 5 (12): e15077. doi:10.1371/journal.pone.0015077.

54. Bruchey, A.K.; Jones, C.E.; Monfils, M.H. (2010). "Fear conditioning by-proxy: social transmission of fear during memory retrieval". Behavioural Brain Research. 214 (1): 80–84. doi:10.1016/j.bbr.2010.04.047. PMC 2975564Freely accessible. PMID 20441779.

55. Jeon, D.; Kim, S.; Chetana, M.; Jo, D.; Ruley, H.E.; Lin, S.Y.; Shin, H.S.; Kinet, Jean-Pierre; Shin, Hee-Sup (2010). "Observational fear learning involves affective pain system and Cav1.2 Ca2+ channels in ACC". Nature Neuroscience. 13 (4): 482–488. doi:10.1038/nn.2504. PMC 2958925Freely accessible. PMID 20190743.

56. Chen, Q.; Panksepp, J.B.; Lahvis, G.P. (2009). "Empathy is moderated by genetic background in mice". PLoS ONE. 4 (2): e4387. doi:10.1371/journal.pone.0004387.

57. Smith, A.V., Proops, L., Grounds, K., Wathan, J. and McComb, K. (2016). "Functionally relevant responses to human facial expressions of emotion in the domestic horse (Equus caballus)". Biology Letters. 12 (2): 20150907. doi:10.1098/rsbl.2015.0907. PMC 4780548Freely accessible. PMID 26864784.

58. Bekoff, Marc (2007). The Emotional Lives of Animals. Novato, California: New World Library. p. 1.

59. (Pg. 1) Bekoff, Marc. The Emotional Lives of Animals. Novato, California: New World Library, 2007.

60. Orlaith, N.F.; Bugnyar, T. (2010). "Do Ravens Show Consolation? Responses to Distressed Others". PLoS ONE. 5 (5): e10605. doi:10.1371/journal.pone.0010605.

61. Edgar, J.L., Paul, E.S. and Nicol, C.J. (2013). "Protective mother hens: Cognitive influences on the avian maternal response". Animal Behaviour. 86 (2): 223–229. doi:10.1016/j.anbehav.2013.05.004. ISSN 0003-3472.

62. Edgar, J.L., Lowe, J.C., Paul, E.S. and Nicol, C.J. (2011). "Avian maternal response to chick distress". Proceedings of the Royal Society B: Biological Sciences. 278 (1721): 3129–3134. doi:10.1098/rspb.2010.2701. ISSN 0962-8452. PMC 3158930Freely accessible. PMID 21389025.

63. Broom, D.M.; Fraser, A.F. (2015). Domestic Animal Behaviour and Welfare (5 ed.). CABI Publishers. pp. 42, 53, 188. ISBN 1780645392.

64. Edgar, J.L., Paul, E.S., Harris, L., Penturn, S. and Nicol, C.J. (2012). "No evidence

for emotional empathy in chickens observing familiar adult conspecifics". PLoS ONE. 7 (2): e31542. doi:10.1371/journal.pone.0031542. PMC 3278448Freely accessible. PMID 22348100.

65. Seligman, M. E. (1972). Learned helplessness. Annual Review of Medicine, 207-412.
66. Seligman, M. E.; Groves, D. P. (1970). "Nontransient learned helplessness". Psychonomic Science. 19: 191–192.
67. Guo, K.; Hall, C.; Hall, S.; Meints, K.; Mills, D. (2007). "Left gaze bias in human infants, rhesus monkeys, and domestic dogs.". Perception. 3. Retrieved 2010-06-24.
68. Alleyne, R. (2008-10-29). "Dogs can read emotion in human faces". Daily Telegraph. London. Retrieved 2010-06-24.
69. Svartberg, K.; Tapper, I.; Temrin, H.; Radesäter, T.; Thorman, S. (2004). "Consistency of personality traits in dogs". Animal Behaviour. 69 (2): 283–291. doi:10.1016/j.anbehav.2004.04.011.
70. Svartberga, K.; Forkman, B. (2002). "Personality traits in the domestic dog (Canis familiaris)". Applied Animal Behaviour Science. 79 (2): 133–155. doi:10.1016/S0168-1591(02)00121-1.
71. Albuquerque, N., Guo, K., Wilkinson, A., Savalli, C., Otta, E. and Mills, D. (2016). "Dogs recognize dog and human emotions". Biology Letters. 12 (1): 20150883. doi:10.1098/rsbl.2015.0883. PMC 4785927Freely accessible. PMID 26763220.
72. Berns, G. S.; Brooks, A. M.; Spivak, M. (2012). "Functional MRI in Awake Unrestrained Dogs". PLoS ONE. 7 (5): e38027. doi:10.1371/journal.pone.0038027. PMC 3350478Freely accessible. PMID 22606363.
73. "Cats do control humans, study finds". LiveScience.com. 2009. Retrieved 2009-07-20.
74. Lion#cite ref-105
75. Fossat, P., Bacqué-Cazenave, J., De Deurwaerdère, P., Delbecque, J.-P. and Cattaert, D. (2014). "Anxiety-like behavior in crayfish is controlled by serotonin". Science. 344 (6189): 1293–1297. doi:10.1126/science.1248811. PMID 24926022.
76. Sneddon, L.U. (2015). "Pain in aquatic animals". Journal of Experimental Biology. 218 (7): 967–976. doi:10.1242/jeb.088823.
77. Fossat, P., Bacqué-Cazenave, J., De Deurwaerdère, P., Cattaert, D. and Delbecque, J.P. (2015). "Serotonin, but not dopamine, controls the stress response and anxiety-like behavior in the crayfish Procambarus clarkii". Journal of Experimental Biology. 218 (17): 2745–2752. doi:10.1242/jeb.120550.

CHAPTER 6

Neuroethology

Chapter Contents

Neuroethology refers to the study of the neural basis of natural behavior in animals. It attempts to understand how sensory organs and central structures process behaviorally relevant stimuli, and how this information is integrated by the central nervous system to produce the behavioral output observed under natural conditions. Many of the concepts and techniques of neuroethology are derived from other biological disciplines, including ethology, neurophysiology, neuroanatomy, neuroendocrinology, and biological cybernetics. A characteristic overall goal of neuroethology is to understand, from mechanistic and evolutionary points of view, both specialization and diversity of neural control among different species.

INTRODUCTION

Neuroethology is the evolutionary and comparative approach to the study of animal behavior and its underlying mechanistic control by the nervous system. This interdisciplinary branch of behavioral neuroscience endeavors to understand how the central nervous system translates biologically relevant stimuli into natural behavior. For example, many bats are capable of echolocation which is used for prey capture and navigation. The auditory system of bats is often cited as an example for how acoustic properties of sounds can be converted into a sensory map of behaviorally relevant features of sounds. Neuroethologists hope to uncover general principles of the nervous system from the study of animals with exaggerated or specialized behaviors.

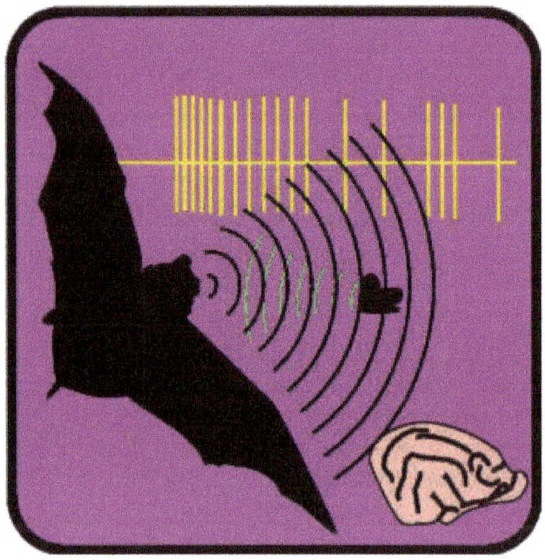

Echolocation in bats is one model system in neuroethology

As its name implies, neuroethology is a multidisciplinary field composed of neurobiology (the study of the nervous system) and ethology (the study of behavior in natural conditions). A central theme of the field of neuroethology, delineating it from other branches of neuroscience, is this focus on natural behavior. Natural behaviors may be thought of as those behaviors generated through means of natural selection (i.e. finding mates, navigation, locomotion, predator avoidance) rather than behaviors in disease states, or behavioral tasks that are particular to the laboratory.

PHILOSOPHY

Neuroethology is an integrative approach to the study of animal behavior that draws upon several disciplines. Its approach stems from the theory that animals' nervous systems have evolved to address problems of sensing and acting in certain environmental niches and that their nervous systems are best understood in the context of the problems they have evolved to solve. In accordance with Krogh's principle, neuroethologists often study animals that are "specialists" in the behavior the researcher wishes to study e.g. honeybees and social behavior, bat echolocation, owl sound localization, etc.

The scope of neuroethological inquiry might be summarized by Jörg-Peter Ewert, a pioneer of neuroethology, when he considers the types of questions central to neuroethology in his 1980 introductory text to the field:

- How are stimuli detected by an organism?
- How are environmental stimuli in the external world represented in the nervous system?
- How is information about a stimulus acquired, stored and recalled by the nervous system?

- How is a behavioral pattern encoded by neural networks?
- How is behavior coordinated and controlled by the nervous system?
- How can the ontogenetic development of behavior be related to neural mechanisms?

Often central to addressing questions in neuroethology are comparative methodologies, drawing upon knowledge about related organisms' nervous systems, anatomies, life histories, behaviors and environmental niches. While it is not unusual for many types of neurobiology experiments to give rise to behavioral questions, many neuroethologists often begin their research programs by observing a species' behavior in its natural environment. Other approaches to understanding nervous systems include the systems identification approach, popular in engineering. The idea is to stimulate the system using a non-natural stimulus with certain properties. The system's response to the stimulus may be used to analyze the operation of the system. Such an approach is useful for linear systems, but the nervous system is notoriously nonlinear, and neuroethologists argue that such an approach is limited. This argument is supported by experiments in the auditory system. These experiments show that neural responses to complex sounds, like social calls, can not be predicted by the knowledge gained from studying the responses due to pure tones (one of the non-natural stimuli favored by auditory neurophysiologists). This is because of the non-linearity of the system.

Modern neuroethology is largely influenced by the research techniques used. Neural approaches are necessarily very diverse, as is evident through the variety of questions asked, measuring techniques used, relationships explored, and model systems employed. Techniques utilized since 1984 include the use of intracellular dyes, which make maps of identified neurons possible, and the use of brain slices, which bring vertebrate brains into better observation through intracellular electrodes (Hoyle 1984). Currently, other fields toward which neuroethology may be headed include computational neuroscience, molecular genetics, neuroendocrinology and epigenetics. The existing field of neural modeling may also expand into neuroethological terrain, due to its practical uses in robotics. In all this, neuroethologists must use the right level of simplicity to effectively guide research towards accomplishing the goals of neuroethology.

Critics of neuroethology might consider it a branch of neuroscience concerned with 'animal trivia'. Though neuroethological subjects tend not to be traditional neurobiological model systems (i.e. *Drosophila*, *C. elegans*, or *Danio rerio*), neuroethological approaches emphasizing comparative methods have uncovered many concepts central to neuroscience as a whole, such as lateral inhibition, coincidence detection, and sensory maps. The discipline of neuroethology has also discovered and explained the only vertebrate behavior for which the entire neural circuit has been described: the electric fish jamming avoidance response. Beyond its conceptual contributions, neuroethology makes indirect contributions to advancing human health. By understanding simpler nervous systems, many clinicians have used concepts uncovered by neuroethology and other branches of neuroscience to develop treatments for devastating human diseases.

HISTORY

The field of neuroethology owes part of its existence to the establishment of ethology as a unique discipline within the discipline of Zoology. Although animal behavior had been studied since the time of Aristotle (384-342 BC), it was not until the early twentieth century that ethology finally became distinguished from natural science (a strictly descriptive field) and ecology. The main catalysts behind this new distinction were the research and writings of Konrad Lorenz and Niko Tinbergen.

Konrad Lorenz was born in Austria in 1903, and is widely known for his contribution of the theory of fixed action patterns (FAPs): endogenous, instinctive behaviors involving a complex sequence of movements that are triggered ("released") by a certain kind of stimulus. This sequence always proceeds to completion, even if the original stimulus is removed. It is also species-specific and performed by nearly all members. Lorenz constructed his famous "hydraulic model" to help illustrate this concept, as well as the concept of action specific energy, or drives.

Niko Tinbergen was born in the Netherlands in 1907 and worked closely with Lorenz in the development of the FAP theory; their studies focused on the egg retrieval response of nesting geese. Tinbergen performed extensive research on the releasing mechanisms of particular FAPs, and used the bill-pecking behavior of baby herring gulls as his model system. This led to the concept of the supernormal stimulus. Tinbergen is also well known for his four questions that he believed ethologists should be asking about any given animal behavior; among these is that of the mechanism of the behavior, on a physiological, neural and molecular level, and this question can be thought of in many regards as the keystone question in neuroethology. Tinbergen also emphasized the need for ethologists and neurophysiologists to work together in their studies, a unity that has become a reality in the field of neuroethology.

Unlike behaviorism, which studied animals' reactions to non-natural stimuli in artificial, laboratory conditions, ethology sought to categorize and analyze the natural behaviors of animals in a field setting. Similarly, neuroethology asks questions about the neural bases of *naturally occurring* behaviors, and seeks to mimic the natural context as much as possible in the laboratory.

Although the development of ethology as a distinct discipline was crucial to the advent of neuroethology, equally important was the development of a more comprehensive understanding of Neuroscience. Contributors to this new understanding were the Spanish Neuroanatomist, Ramon y Cajal (born in 1852), and physiologists Charles Sherrington, Edgar Adrian, Alan Hodgkin, and Andrew Huxley. Charles Sherrington, who was born in Great Britain in 1857, is famous for his work on the nerve synapse as the site of transmission of nerve impulses, and for his work on reflexes in the spinal cord. His research also led him to hypothesize that every muscular activation is coupled to an inhibition of the opposing muscle. He was awarded a Nobel Prize for his work in 1932 along with Lord Edgar Adrian who made the first physiological recordings of neural activity from single nerve fibers.

Alan Hodgkin and Andrew Huxley (born 1914 and 1917, respectively, in Great Britain), are known for their collaborative effort to understand the production of action potentials in giant squid neurons. The pair also proposed the existence of ion channels to facilitate action potential initiation, and were awarded the Nobel Prize in 1963 for their efforts.

As a result of this pioneering research, many scientists then sought to connect the physiological aspects of the nervous and sensory systems to specific behaviors. These scientists – Karl von Frisch, Erich von Holst, and Theodore Bullock – are frequently referred to as the "fathers" of neuroethology. Neuroethology did not really come into its own, though, until the 1970s and 1980s, when new, sophisticated experimental methods allowed researchers such as Masakazu Konishi, Walter Heiligenberg, Jörg-Peter Ewert, and others to study the neural circuits underlying verifiable behavior.

MODERN NEUROETHOLOGY

The International Society for Neuroethology represents the present discipline of neuroethology, which was founded on the occasion of the NATO-Advanced Study Institute "Advances in Vertebrate Neuroethology" (August 13–24, 1981) organized by J.-P. Ewert, D.J. Ingle and R.R. Capranica, held at the University of Kassel in Hofgeismar, Germany (cf. report Trends in Neurosci. 5:141-143,1982). Its first president was Theodore H. Bullock. The society has met every three years since its first meeting in Tokyo in 1986.

Its membership draws from many research programs around the world; many of its members are students and faculty members from medical schools and neurobiology departments from various universities. Modern advances in neurophysiology techniques have enabled more exacting approaches in an ever-increasing number of animal systems, as size limitations are being dramatically overcome. Survey of the most recent (2007) congress of the ISN meeting symposia topics gives some idea of the field's breadth:

- Comparative aspects of spatial memory (rodents, birds, humans, bats)
- Influences of higher processing centers in active sensing (primates, owls, electric fish, rodents, frogs)
- Animal signaling plasticity over many time scales (electric fish, frogs, birds)
- Song production and learning in passerine birds
- Primate sociality
- Optimal function of sensory systems (flies, moths, frogs, fish)
- Neuronal complexity in behavior (insects, computational)
- Contributions of genes to behavior (Drosophila, honeybees, zebrafish)
- Eye and head movement (crustaceans, humans, robots)
- Hormonal actions in brain and behavior (rodents, primates, fish, frogs, and birds)
- Cognition in insects (honeybee)

APPLICATION TO TECHNOLOGY

Neuroethology can help create advancements in technology through an advanced understanding of animal behavior. Model systems were generalized from the study of simple and related animals to humans. For example, the neuronal cortical space map discovered in bats, a specialized champion of hearing and navigating, elucidated the concept of a computational space map. In addition, the discovery of the space map in the barn owl led to the first neuronal example of the Jeffress model. This understanding is translatable to understanding spatial localization in humans, a mammalian relative of the bat. Today, knowledge learned from neuroethology are being applied in new technologies. For example, Randall Beer and his colleagues used algorithms learned from insect walking behavior to create robots designed to walk on uneven surfaces (Beer et al.). Neuroethology and technology contribute to one another bidirectionally.

Neuroethologists seek to understand the neural basis of a behavior as it would occur in an animal's natural environment but the techniques for neurophysiological analysis are lab-based, and cannot be performed in the field setting. This dichotomy between field and lab studies poses a challenge for neuroethology. From the neurophysiology perspective, experiments must be designed for controls and objective rigor, which contrasts with the ethology perspective—that the experiment be applicable to the animal's natural condition, which is uncontrolled, or subject to the dynamics of the environment. An early example of this is when Walter Rudolf Hess developed focal brain stimulation technique to examine a cat's brain controls of vegetative functions in addition to other behaviors. Even though this was a breakthrough in technological abilities and technique, it was not used by many neuroethologists originally because it compromised a cat's natural state, and, therefore, in their minds, devalued the experiments' relevance to real situations.

When intellectual obstacles like this were overcome, it led to a golden age of neuroethology, by focusing on simple and robust forms of behavior, and by applying modern neurobiological methods to explore the entire chain of sensory and neural mechanisms underlying these behaviors (Zupanc 2004). New technology allows neuroethologists to attach electrodes to even very sensitive parts of an animal such as its brain while it interacts with its environment. The founders of neuroethology ushered this understanding and incorporated technology and creative experimental design. Since then even indirect technological advancements such as battery-powered and waterproofed instruments have allowed neuroethologists to mimic natural conditions in the lab while they study behaviors objectively. In addition, the electronics required for amplifying neural signals and for transmitting them over a certain distance have enabled neuroscientists to record from behaving animals performing activities in naturalistic environments. Emerging technologies can complement neuroethology, augmenting the feasibility of this valuable perspective of natural neurophysiology. Another challenge, and perhaps part of the beauty of neuroethology, is experimental design. The value of neuroethological criteria speak to the reliability of these experiments, because these discoveries represent behavior in the environments in which they evolved. Neuroethologists foresee future advancements through using new technologies and techniques, such as computational neuroscience, neuroendocrinology, and molecular genetics that mimic natural environments.

CASE STUDIES

Jamming Avoidance Response

In 1963, Akira Watanabe and Kimihisa Takeda discovered the behavior of the jamming avoidance response in the knifefish *Eigenmannia* sp. In collaboration with T.H. Bullock and colleagues, the behavior was further developed. Finally, the work of W. Heiligenberg expanded it into a full neuroethology study by examining the series of neural connections that led to the behavior. *Eigenmannia* is a weakly electric fish that can generate electric discharges through electrocytes in its tail. Furthermore, it has the ability to electrolocate by analyzing the perturbations in its electric field. However, when the frequency of a neighboring fish's current is very close (less than 20 Hz difference) to that of its own, the fish will avoid having their signals interfere through a behavior known as Jamming Avoidance Response. If the neighbor's frequency is higher than the fish's discharge frequency, the fish will lower its frequency, and vice versa. The sign of the frequency difference is determined by analyzing the "beat" pattern of the incoming interference which consists of the combination of the two fish's discharge patterns.

Neuroethologists performed several experiments under *Eigenmannia*'s natural conditions to study how it determined the sign of the frequency difference. They manipulated the fish's discharge by injecting it with curare which prevented its natural electric organ from discharging. Then, an electrode was placed in its mouth and another was placed at the tip of its tail. Likewise, the neighboring fish's electric field was mimicked using another set of electrodes. This experiment allowed neuroethologists to manipulate different discharge frequencies and observe the fish's behavior. From the results, they were able to conclude that the electric field frequency, rather than an internal frequency measure, was used as a reference. This experiment is significant in that not only does it reveal a crucial neural mechanism underlying the behavior but also demonstrates the value neuroethologists place on studying animals in their natural habitats.

Feature Analysis in Toad Vision

The recognition of prey and predators in the toad was first studied in depth by Jörg-Peter Ewert (Ewert 1974; see also 2004). He began by observing the natural prey-catching behavior of the common toad (*Bufo bufo*) and concluded that the animal followed a sequence that consisted of stalking, binocular fixation, snapping, swallowing and mouth-wiping. However, initially, the toad's actions were dependent on specific features of the sensory stimulus: whether it demonstrated worm or anti-worm configurations. It was observed that the worm configuration, which signaled prey, was initiated by movement along the object's long axis, whereas anti-worm configuration, which signaled predator, was due to movement along the short axis. (Zupanc 2004).

Ewert and coworkers adopted a variety of methods to study the predator versus prey behavior response. They conducted recording experiments where they inserted electrodes into the brain, while the toad was presented with worm or anti-worm stimuli. This technique was repeated at different levels of the visual system and also allowed feature detectors to be identified. In focus was the discovery of prey-selective neurons in the optic tectum, whose axons could be traced towards the snapping pattern generating cells in the hypoglossal nucleus. The discharge patterns of prey-selective tectal neurons in response to prey objects – in freely moving toads –predicted" prey-catching reactions such as snapping. Another approach, called stimulation experiment, was carried out in freely moving toads. Focal electrical stimuli were applied to different regions of the brain, and the toad's response was observed. When the thalamic-pretectal region was stimulated, the toad exhibited escape responses, but when the tectum was stimulated in an area close to prey-selective neurons, the toad engaged in prey catching behavior (Carew 2000). Furthermore, neuroanatomical experiments were carried out where the toad's thalamic-pretectal/tectal connection was lesioned and the resulting deficit noted: the prey-selective properties were abolished both in the responses of prey-selective neurons and in the prey catching behavior. These and other experiments suggest that prey selectivity results from pretecto-tectal influences.

Ewert and coworkers showed in toads that there are stimulus-response mediating pathways that translate perception (of visual sign stimuli) into action (adequate behavioral responses). In addition there are modulatory loops that initiate, modify or specify this mediation (Ewert 2004). Regarding the latter, for example, the telencephalic caudal ventral striatum is involved in a loop gating the stimulus-response mediation in a manner of directed attention. The telencephalic ventral medial pallium („primordium hippocampi"), however, is involved in loops that either modify prey-selection due to associative learning or specify prey-selection due to non-associative learning, respectively.

COMPUTATIONAL NEUROETHOLOGY

Computational neuroethology (CN or CNE) is concerned with the computer modelling of the neural mechanisms underlying animal behaviors. Computational neuroethology was first argued for in depth by Randall Beer and by Dave Cliff both of whom acknowledged the strong influence of Michael Arbib's *Rana Computatrix* computational model of neural mechanisms for visual guidance in frogs and toads.

CNE systems work within a closed-loop environment; that is, they perceive their (perhaps artificial) environment directly, rather than through human input, as is typical in AI systems. For example, Barlow et al. developed a time-dependent model for the retina of the horseshoe crab *Limulus polyphemus* on a Connection Machine (Model CM-2). Instead of feeding the model retina with idealized input signals, they exposed the simulation to digitized video sequences made underwater, and compared its response with those of real animals.

MODEL SYSTEMS

- Bat echolocation – nocturnal flight navigation and prey capture; location of objects using echo returns of its own call
- Oscine bird song – zebra finch (*Taeniopygia guttata*), canary (*Serinus canaria*) and white-crowned sparrow (*Zonotrichia leucophrys*); song learning as a model for human speech development
- Electric fish – navigation, communication, Jamming Avoidance Response (JAR), corollary discharge, expectation generators, and spike timing dependent plasticity

- Barn owl auditory spatial map – nocturnal prey location and capture
- Toad vision – discrimination of prey versus predator – Video "Image processing in the toad's visual system: behavior, brain function, artificial neuronal net
- Circadian rhythm – influence of various circadian controlled behaviors by the suprachiasmatic nucleus
- Cricket song – mate attraction and corollary discharge
- Fish Mauthner cells – C-start escape response and underwater directional hearing
- Fly – Microscale directional hearing in *Ormia ochracea*, sex differences of the visual system in Bibionidae, and spatial navigation in chasing behavior of Fannia canicularis
- Noctuid moths – ultrasound avoidance response to bat calls
- Aplysia – learning and memory in startle response
- Rat – spatial memory and navigation
- Salmon homing – olfactory imprinting and thyroid hormones
- Crayfish – escape and startle behaviors, aggression and formation of social hierarchies
- Cichlid fish – aggression and attack behaviors
- Honey bee – learning, navigation, vision, olfaction, flight, aggression, foraging
- Monarch butterfly – navigational mechanisms
- More Model Systems and Information

ANIMAL ECHOLOCATION

Echolocation, also called bio sonar, is the biological sonar used by several kinds of animals. Echolocating animals emit calls out to the environment and listen to the echoes of those calls that return from various objects near them. They use these echoes to locate and identify the objects. Echolocation is used for navigation and for foraging (or hunting) in various environments. Some blind humans have learned to find their way using clicks produced by a device or by mouth.

Echolocating animals include some mammals and a few birds; most notably microchiropteran bats and odontocetes (toothed whales and dolphins), but also in simpler form in other groups such as shrews, one genus of megachiropteran bats (*Rousettus*) and two cave dwelling bird groups, the so-called cave swiftlets in the genus *Aerodramus* (formerly *Collocalia*) and the unrelated Oilbird *Steatornis caripensis*.

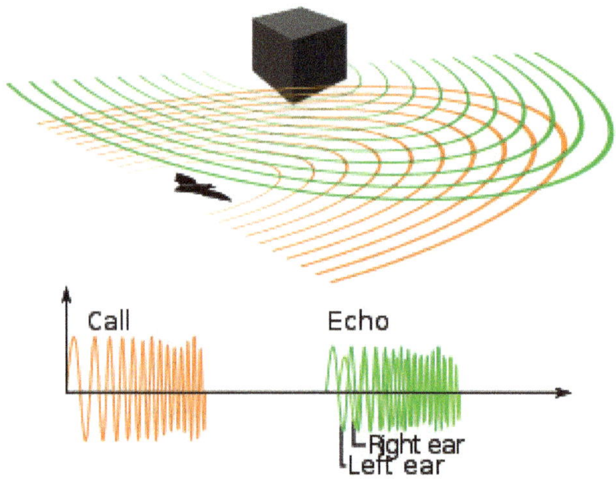

A depiction of the ultrasound signals emitted by a bat, and the echo from a nearby object.

Early Research

The term *echolocation* was coined by Donald Griffin, whose work with Robert Galambos was the first to conclusively demonstrate its existence in bats in 1938. As Griffin described in his book, the 18th century Italian scientist Lazzaro Spallanzani had, by means of a series of elaborate experiments, concluded that bats navigate by hearing and not by vision. In 1912, Sir Hiram Maxim had independently proposed that bats used sound below the human auditory range to avoid obstacles. In 1920, the English physiologist Hartridge proposed instead that bats used frequencies above the range of human hearing, which ultimately proved to be the case

Echolocation in odontocetes was not properly described until two decades after Griffin and Galambos' work, by Schevill and McBride.

Principle

Echolocation is the same as active sonar, using sounds made by the animal itself. Ranging is done by measuring the time delay between the animal's own sound emission and any echoes that return from the environment. The relative intensity of sound received at each ear as well as the time delay between arrival at the two ears provide information about the horizontal angle (azimuth) from which the reflected sound waves arrive.

Unlike some human-made sonars that rely on many extremely narrow beams and many receivers to localize a target (multibeam sonar), animal echolocation has only one transmitter and two receivers (the ears). Echolocating animals have two ears positioned slightly apart. The echoes returning to the two ears arrive at different times and at different loudness levels, depending on the position of the object generating the echoes. The time and loudness differences are used by the animals to perceive distance and direction. With echolocation, the bat or other animal can see not only where it is going but also how big another animal is, what kind of animal it is, and other features..

At the most basic level, echolocation is based on the neuro anatomy of auditory brain circuitry. In essence, ascending brain pathways in the brain stem allow the brain to calculate the difference between the two ears to very small fractions of a second.

Bats

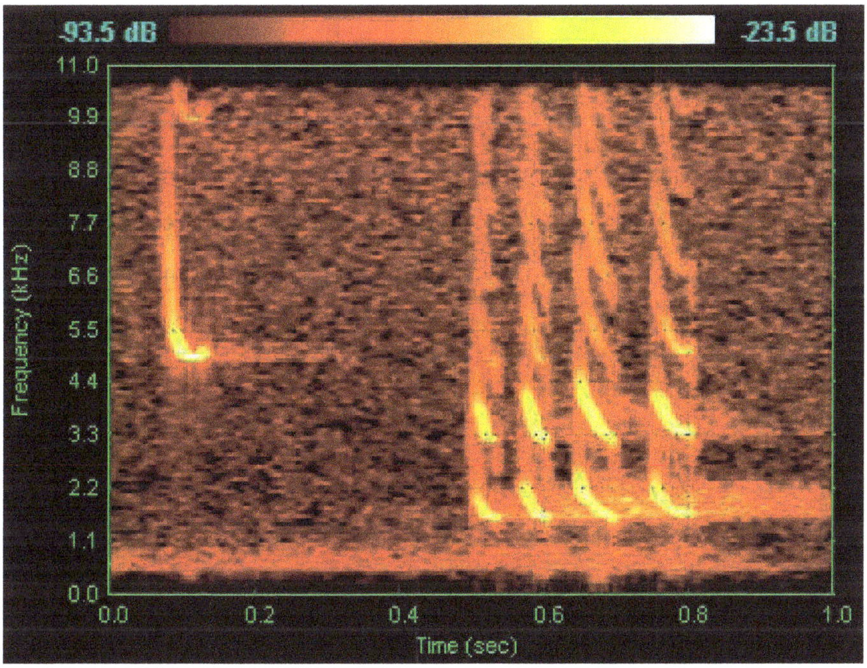

Spectrogram of Pipistrellus Bat vocalizations. Detail is shown as the pulses transition to a faster repetition rate. The bat appears to use a hybrid pulse which combines a sharp falling frequency chirp with an extended constant frequency tail. Such a waveform may offer combined benefits of range estimation as well as Doppler shift detection.

Microbats use echolocation to navigate and forage, often in total darkness. They generally emerge from their roosts in caves, attics, or trees at dusk and hunt for insects into the night. Their use of echolocation allows them to occupy a niche where there are often many insects (that come out at night since there are fewer predators then), less competition for food, and fewer species that may prey on the bats themselves.

Microbats generate ultrasound via the larynx and emit the sound through the open mouth or, much more rarely, the nose. The latter is most pronounced in the horseshoe bats (*Rhinolophus spp.*). Microbat 🔊 calls (help·info)range in frequency from 14,000 to well over 100,000 Hz, mostly beyond the range of the human ear (typical human hearing range is considered to be from 20 Hz to 20,000 Hz). Bats may estimate the elevation of targets by interpreting the interference patterns caused by the echoes reflecting from the tragus, a flap of skin in the external ear. There are two hypotheses about the evolution of echolocation in bats. The first suggests that laryngeal echolocation evolved twice in Chiroptera, once in the Yangochiroptera and once in the Horseshoe bats (Rhinolophidae). The second proposes that laryngeal echolocation had a single origin in Chiroptera, was subsequently lost in the family Pteropodidae (all megabats), and later evolved as a system of tongue-clicking in the genus *Rousettus*.

Individual bat species echolocate within specific frequency ranges that suit their environment and prey types. This has sometimes been used by researchers to identify bats flying in an area simply by recording their calls with ultrasonic recorders known as "bat detectors". However echolocation calls are not always species specific and some bats overlap in the type of calls they use so recordings of echolocation calls cannot be used to identify all bats. In recent years researchers in several countries have developed "bat call libraries" that contain recordings of local bat species that have been identified known as "reference calls" to assist with identification.

Since the 1970s there has been an ongoing controversy among researchers as to whether bats use a form of processing known from radar termed coherent cross-correlation. Coherence means that the phase of the echolocation signals is used by the bats, while cross-correlation just implies that the outgoing signal is compared with the returning echoes in a running process. Today most - but not all - researchers believe that they use cross-correlation, but in an incoherent form, termed a filter bank receiver.

When searching for prey they produce sounds at a low rate (10-20 clicks/second). During the search phase the sound emission is coupled to respiration, which is again coupled to the wingbeat. This coupling appears to dramatically conserve energy as there is little to no additional energetic cost of echolocation to flying bats. After detecting a potential prey item, microbats increase the rate of pulses, ending with the **terminal buzz**, at rates as high as 200 clicks/second. During approach to a detected target, the duration of the sounds is gradually decreased, as is the energy of the sound.

Calls and Ecology

Bats belonging to the suborder Microchiroptera (microbats) occupy a diverse set of ecological conditions - they can be found living in environments as different as Europe and Madagascar, and hunting for food sources as different as insects, frogs, nectar, fruit, and blood. Additionally, the characteristics of an echolocation call are adapted to the particular environment, hunting behavior, and food source of the particular bat. However, this adaptation of echolocation calls to ecological factors is constrained by the phylogenetic relationship of the bats, leading to a process known as descent with modification, and resulting in the diversity of the Microchiroptera today.

Acoustic Features

Describing the diversity of bat echolocation calls requires examination of the frequency and temporal features of the calls. It is the variations in these aspects that produce echolocation calls suited for different acoustic environments and hunting behaviors.

- **Frequency**: Bat call frequencies range from as low as 11 kHz to as high as 212 kHz. Insectivorous aerial-hawking bats have a call frequency between 20 kHz and 60 kHz because it is the frequency that gives the best range and image acuity and makes them less conspicuous to insects. However, low frequencies are adaptive for some species with different prey and environments. *Euderma maculatum*, a species that feeds on moths, uses a particularly low frequency of 12.7 kHz that cannot be heard by moths.
- **Frequency modulation and constant frequency**: Echolocation calls can be composed of two different types of frequency structures: frequency modulated (FM) sweeps, and constant frequency (CF) tones. A particular call can consist of one, the other, or both structures. An FM sweep is a broadband signal – that is, it contains a downward sweep through a range of frequencies. A CF tone is a narrowband signal: the sound stays constant at one frequency throughout its duration.
- **Intensity**: Echolocation calls have been measured at intensities anywhere between 60 and 140 decibels. Certain microbat species can modify their call intensity mid-call, lowering the intensity as they approach objects that reflect sound strongly. This prevents the returning echo from deafening the bat. High-intensity calls such as those from aerial-hawking bats (133 dB) are adaptive to hunting in open skies. Their high intensity calls are necessary to even have moderate detection of surroundings because air has a high absorption of ultrasound and because insects' size only provide a small target for sound reflection. Additionally, the so-called "whispering bats" have adapted low-amplitude echolocation so that their prey, moths, which are able to hear echolocation calls, are less able to detect and avoid an oncoming bat.
- **Harmonic composition**: Calls can be composed of one frequency or multiple frequencies comprising a harmonic series. In the latter case, the call is usually dominated by a certain harmonic ("dominant" frequencies are those present at higher intensities than other harmonics present in the call).
- **Call duration**: A single echolocation call (a call being a single continuous trace on a sound spectrogram, and a series of calls comprising a sequence or pass) can last anywhere from 0.2 to 100 milliseconds in duration, depending on the stage of prey-catching behavior that the bat is engaged in. For example, the duration of a call usually decreases when the bat is in the final stages of prey capture – this enables the bat to call more rapidly without overlap of call and echo. Reducing duration comes at the cost of having less total sound available for reflecting off objects and being heard by the bat.
- **Pulse interval**: The time interval between subsequent echolocation calls (or pulses) determines two aspects of a bat's perception. First, it establishes how quickly the bat's auditory scene information is updated. For example, bats increase the repetition rate of their calls (that is, decrease the pulse interval) as they home in on a target. This allows the bat to get new information regarding the target's location at a faster rate when it needs it most. Secondly, the pulse interval determines the maximum range that bats can detect objects. This is because bats can only keep track of the echoes from one call at a time; as soon

as they make another call they stop listening for echoes from the previously made call. For example, a pulse interval of 100 ms (typical of a bat searching for insects) allows sound to travel in air roughly 34 meters so a bat can only detect objects as far away as 17 meters (the sound has to travel out and back). With a pulse interval of 5 ms (typical of a bat in the final moments of a capture attempt), the bat can only detect objects up to 85 cm away. Therefore, the bat constantly has to make a choice between getting new information updated quickly and detecting objects far away.

FM Signal Advantages

Echolocation call produced by *Pipistrellus pipistrellus*, an FM bat. The ultrasonic call has been "heterodyned" - multiplied by a constant frequency to produce frequency subtraction, and thus an audible sound - by a bat detector. A key feature of the recording is the increase in the repetition rate of the call as the bat nears its target - this is called the "terminal buzz".

The major advantage conferred by an FM signal is extremely precise range discrimination, or localization, of the target. J.A. Simmons demonstrated this effect with a series of elegant experiments that showed how bats using FM signals could distinguish between two separate targets even when the targets were less than half a millimeter apart. This ability is due to the broadband sweep of the signal, which allows for better resolution of the time delay between the call and the returning echo, thereby improving the cross correlation of the two. Additionally, if harmonic frequencies are added to the FM signal, then this localization becomes even more precise.

One possible disadvantage of the FM signal is a decreased operational range of the call. Because the energy of the call is spread out among many frequencies, the distance at which the FM-bat can detect targets is limited. This is in part because any echo returning at a particular frequency can only be evaluated for a brief fraction of a millisecond, as the fast downward sweep of the call does not remain at any one frequency for long.

CF Signal Advantages

The structure of a CF signal is adaptive in that it allows the CF-bat to detect both the velocity of a target, and the fluttering of a target's wings as Doppler shifted frequencies. A Doppler shift is an alteration in sound wave frequency, and is produced in two relevant situations: when the bat and its target are moving relative to each other, and when the target's wings are oscillating back and forth. CF-bats must compensate for Doppler shifts, lowering the frequency of their call in response to echoes of elevated frequency - this ensures that the returning echo remains at the frequency to which the ears of the bat are most finely tuned. The oscillation of a target's wings also produces amplitude shifts, which gives a CF-bat additional help in distinguishing a flying target from a stationary one. (Schnitzler and Flieger 1983; Zupanc 2004; Simmons and Stein 1980; Grinnell 1995; Neuweiler 2003; Jones and Teeling 2006)

Additionally, because the signal energy of a CF call is concentrated into a narrow frequency band, the operational range of the call is much greater than that of an FM signal. This relies on the fact that echoes returning within the narrow frequency band can be summed over the entire length of the call, which maintains a constant frequency for up to 100 milliseconds.

Acoustic Environments of FM and CF Signals

- **FM**: An FM (frequency modulated) component is excellent for hunting prey while flying in close, cluttered environments. Two aspects of the FM signal account for this fact: the precise target localization conferred by the broadband signal, and the short duration of the call. The first of these is essential because in a cluttered environment, the bats must be able to resolve their prey from large amounts of background noise. The 3D localization abilities of the broadband signal enable the bat to do exactly that, providing it with what Simmons and Stein (1980) call a "clutter rejection strategy." This strategy is further improved by the use of harmonics, which, as previously stated, enhance the localization properties of the call. The short duration of the FM call is also best in close, cluttered environments because it enables the bat to emit many calls extremely rapidly without overlap. This means that the bat can get an almost continuous stream of information – essential when objects are close, because they will pass by quickly – without confusing which echo corresponds to which call. (Neuweiler 2003; Simmons and Stein 1980; Jones and Teeling 2006; Fenton 1995)
- **CF**: A CF (constant frequency) component is often used by bats hunting for prey while flying in open, clutter-free environments, or by bats that wait on perches for their prey to appear. The success of the former strategy is due to two aspects of the CF call, both of which confer excellent prey-detection abilities. First, the greater working range of the call allows bats to detect targets present at great distances – a common situation in open environments. Second, the length of the call is also suited for targets at great distances: in this case, there is a decreased chance that the long call will overlap with the returning echo. The latter strategy is made possible by the fact that the long, narrowband call allows the bat to detect Doppler shifts, which would be produced by an insect moving either towards or away from a perched bat. (Neuweiler 2003; Simmons and Stein 1980; Jones and Teeling 2006; Fenton 1995)

Neural Mechanisms in the Brain

Because bats use echolocation to orient themselves and to locate objects, their auditory systems are adapted for this purpose, highly specialized for sensing and interpreting the stereotyped echolocation calls characteristic of their own species. This specialization is evident from the inner ear up to the highest levels of information processing in the auditory cortex.

Inner Ear and Primary Sensory Neurons

Both CF and FM bats have specialized inner ears which allow them to hear sounds in the ultrasonic range, far outside the range of human hearing. Although in most other aspects, the bat's auditory organs are similar to those of most other mammals, certain bats (horseshoe bats, *Rhinolophus spp.* and the moustached bat, *Pteronotus parnelii*) with a constant frequency (CF) component to their call (known as high duty cycle bats) do have a few additional adaptations for detecting the predominant frequency (and harmonics) of the CF vocalization. These include a narrow frequency "tuning" of the inner ear organs, with an especially large area responding to the frequency of the bat's returning echoes (Neuweiler 2003).

The basilar membrane within the cochlea contains the first of these specializations for echo information processing. In bats that use CF signals, the section of membrane that responds to the frequency of returning echoes is much larger than the region of response for any other frequency. For example, in the horseshoe bat, there is a disproportionately lengthened and thickened section of the membrane that responds to sounds around 83 kHz, the constant frequency of the echo produced by the bat's call. This area of high sensitivity to a specific, narrow range of frequency is known as anacoustic fovea.

Odontocetes (toothed whales and dolphins) have similar cochlear specializations to those found in bats. Odontocetes also have the highest neural investment of any cochleae reported to date with ratios of greater than 1500 ganglion cells/mm of basilar membrane.

Further along the auditory pathway, the movement of the basilar membrane results in the stimulation of primary auditory neurons. Many of these neurons are specificallytuned (respond most strongly) to the narrow frequency range of returning echoes of CF calls. Because of the large size of the acoustic fovea, the number of neurons responding to this region, and thus to the echo frequency, is especially high.

Inferior Colliculus

In the Inferior colliculus, a structure in the bat's midbrain, information from lower in the auditory processing pathway is integrated and sent on to the auditory cortex. As George Pollak and others showed in a series of papers in 1977, the interneurons in this region have a very high level of sensitivity to time differences, since the time delay between a call and the returning echo tells the bat its distance from the target object. Especially interesting is that while most neurons respond more quickly to stronger stimuli, collicular neurons maintain their timing accuracy even as signal intensity changes.

These interneurons are specialized for time sensitivity in several ways. First, when activated, they generally respond with only one or two action potentials. This short duration of response allows their action potentials to give a very specific indication of the exact moment of the time when the stimulus arrived, and to respond accurately to stimuli that occur close in time to one another. In addition, the neurons have a very low threshold of activation – they respond quickly even to weak stimuli. Finally, for FM signals, each interneuron is tuned to a specific frequency within the sweep, as well as to that same frequency in the following echo. There is specialization for the CF component of the call at this level as well. The high proportion of neurons responding to the frequency of the acoustic fovea actually increases at this level.

Auditory Cortex

The auditory cortex in bats is quite large in comparison with other mammals. Various characteristics of sound are processed by different regions of the cortex, each providing different information about the location or movement of a target object. Most of the existing studies on information processing in the auditory cortex of the bat have been done by Nobuo Suga on the mustached bat, *Pteronotus parnellii*. This bat's call has both CF tone and FM sweep components.

Suga and his colleagues have shown that the cortex contains a series of "maps" of auditory information, each of which is organized systematically based on characteristics of sound such as frequency and amplitude. The neurons in these areas respond only to a specific combination of frequency and timing (sound-echo delay), and are known as combination-sensitive neurons.

The systematically organized maps in the auditory cortex respond to various aspects of the echo signal, such as its delay and its velocity. These regions are composed ofcombination sensitive neurons that require at least two specific stimuli to elicit a response. The neurons vary systematically across the maps, which are organized by acoustic features of the sound and can be two dimensional. The different features of the call and its echo are used by the bat to determine important characteristics of their prey. The maps include:

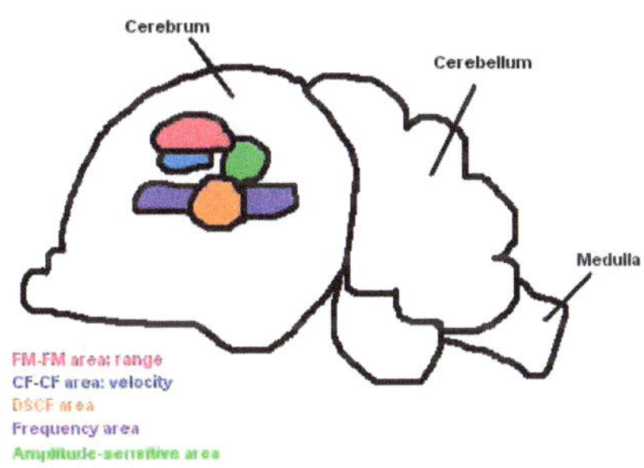

Sketch of the regions of the auditory cortex in a bat's brain

- **FM-FM area**: This region of the cortex contains FM-FM combination-sensitive neurons. These cells respond only to the combination of two FM sweeps: a call and its echo. The neurons in the FM-FM region are often referred to as "delay-tuned," since each responds to a specific time delay between the original call and the echo, in order to find the distance from the target object (the range). Each neuron also shows specificity for one harmonic in the original call and a different harmonic in the echo. The neurons within the FM-FM area of the cortex of *Pteronotus* are organized into columns, in which the delay time is constant vertically but increases across the horizontal plane. The result is that range is encoded by location on the cortex, and increases systematically across the FM-FM area.

- **CF-CF area**: Another kind of combination-sensitive neuron is the CF-CF neuron. These respond best to the combination of a CF call containing two given frequencies – a call at 30 kHz (CF1) and one of its additional harmonics around 60 or 90 kHz (CF2 or CF3) – and the corresponding echoes. Thus, within the CF-CF region, the changes in echo frequency caused by the Doppler shift can be compared to the frequency of the original call to calculate the bat's velocity relative to its target object. As in the FM-FM area, information is encoded by its location within the map-like organization of the region. The CF-CF area is first split into the distinct CF1-CF2 and CF1-CF3 areas. Within each area, the CF1 frequency is organized on an axis, perpendicular to the CF2 or CF3 frequency axis. In the resulting grid, each neuron codes for a certain combination of frequencies that is indicative of a specific velocity
- **DSCF area**: This large section of the cortex is a map of the acoustic fovea, organized by frequency and by amplitude. Neurons in this region respond to CF signals that have been Doppler shifted (in other words, echoes only) and are within the same narrow frequency range to which the acoustic fovea responds. For *Pteronotus*, this is around 61 kHz. This area is organized into columns, which are arranged radially based on frequency. Within a column, each neuron responds to a specific combination of frequency and amplitude. Suga's studies have indicated that this brain region is necessary for frequency discrimination.

Toothed Whales

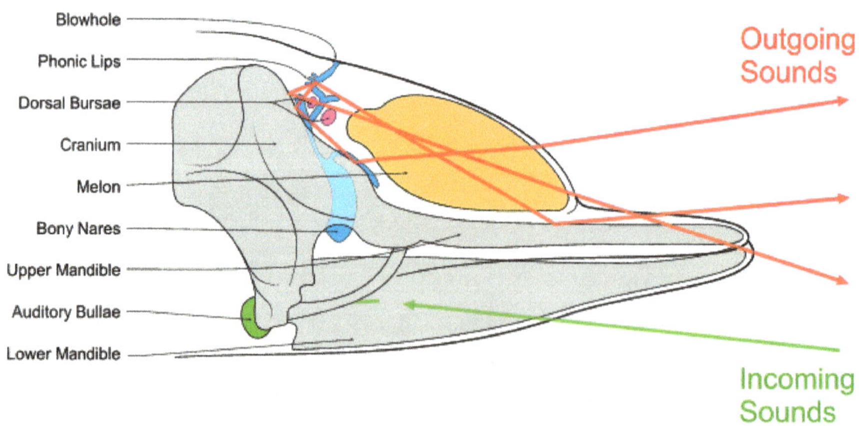

Diagram illustrating sound generation, propagation and reception in a toothed whale. Outgoing sounds are red and incoming ones are green

Biosonar is valuable to toothed whales (suborder Odontoceti), including dolphins, porpoises, river dolphins, killer whales and sperm whales, because they live in an underwater habitat that has favourable acoustic characteristics and where vision is extremely limited in range due to absorption or turbidity.

Cetacean evolution consisted of three main radiations. Throughout the middle and late Eocene periods (49-31.5 million years ago), archaeocetes, primitive toothed Cetacea that arose from terrestrial mammals with the creation of aquatic adaptations, were the only known archaic Cetacea. These primitive aquatic mammals did not possess the ability to echolocate, although they did have slightly adapted underwater hearing. The morphology of acoustically isolated ear bones in basilosaurid archaeocetes indicates that this order had directional hearing underwater at low to mid frequencies by the late middle Eocene.However, with the extinction of archaeocete at the onset of the Oligocene, two new lineages in the early Oligocene period (31.5-28 million years ago) comprised a second radiation. These early mysticetes (baleen whales) and odontocetes can be dated back to the middle Oligocene in New Zealand. Based on past phylogenies, it has been found that the evolution of odontocetes is monophyletic, suggesting that echolocation evolved only once 36 to 34 million years ago. Dispersal rates routes of early odontocetes included transoceanic travel to new adaptive zones. The third radiation occurred later in the Neogene, when present dolphins and their relatives evolved to be the most common species in the modern sea.

The evolution of echolocation could be attributed to several theories. There are two proposed drives for the hypotheses of cetacean radiation, one biotic and the other abiotic in nature. The first, adaptive radiation, is the result of a rapid divergence into new adaptive zones. This results in diverse, ecologically different clades that are incomparable. Clade Neocete (crown cetacean) has been characterized by an evolution from archaeocetes and a dispersion across the world's oceans, and even estuaries and rivers. These ecological opportunities were the result of abundant dietary resources with low competition for hunting. This hypothesis of lineage diversification, however, can be unconvincing due to a lack of support for rapid speciation early in cetacean history. A second, more abiotic drive is better supported. Physical restructuring of the oceans has played a role in echolocation radiation. This was a result of global climate change at the Eocene-Oligocene boundary; from a greenhouse to an icehouse world. Tectonic openings created the emergence of the Southern ocean with a free flowing Antarctic Circumpolar current. These events allowed for a selection regime characterized by the ability to locate and capture prey in turbid river waters, or allow odontocetes to invade and feed at depths below the photic zone. Further studies have found that echolocation below the photic zone could have been a predation adaptation to diel migrating cephalopods. Since its advent, there has been adaptive radiation especially in the family Delphinidae (dolphins) in which echolocation has become extremely derived.

Two proteins have been found to play a major role in toothed whale echolocation. Prestin, a motor protein of the outer hair cells of the inner ear of the mammalian cochlea, has an association between the number of nonsynonymous substitutions and hearing sensitivity. It has undergone two clear episodes of accelerated protein evolution in cetaceans: on the ancestral branch of odontocetes and on the branch leading to delphinioidae. The first episode of acceleration is connected to odontocete divergence, when echolocation first developed, and the second occurs with the increase in echolocation frequency seen in the delphinioidae family. Cldn14, a member of the tight junction proteins which form barriers between inner ear cells, shows exactly the same evolutionary pattern as Prestin. The two events of protein evolution, for Prestin and Cldn14, occurred at the same times as the tectonic opening of the Drake Passage (34-31 Ma) and the Antarctic ice growth at the middle Miocene climate transition (14 Ma), with the divergence of odontocetes and mysticetes occurring with the former, and the speciation of delphinioidae with the latter. There is a strong connection between these proteins, the ocean restructuring events, and the echolocation evolution.

One specific type of echolocation, narrow-band high frequency (NBHF) clicks, evolved at least four times in groups of odontocetes, including the pygmy sperm whale (Kogiidae) and porpoise (Phocoenidae) families, *Pontoporia blainvillei*, the genus *Cephalorhynchus*, and part of the genus *Lagenorhynchus*. These high frequency clicks likely evolved as adaptation of predator avoidance, as they inhabit areas that have many killer whales and the signals are inaudible to killer whales due to the absence of energy below 100 kHz.

Another reason for variation in echolocation is habitat. For all sonar systems the limiting factor deciding whether a returning echo is detected is the echo-to-noise ratio (ENR). The ENR is given by the emitted source level (SL) plus the target strength, minus the two-way transmission loss (absorption and spreading) and the received noise. Animals will adapt either to maximize range under noise-limited conditions (increase source level) or to reduce noise clutter in a shallow and/or littered habitat (decrease source level). In cluttered habitats, such as coastal areas, prey ranges are smaller and species like the Commerson's dolphin (Cephalorhynchus commersonii) will have lowered source levels to better suit their environment

Toothed whales emit a focused beam of high-frequency clicks in the direction that their head is pointing. Sounds are generated by passing air from the bony nares through the phonic lips. These sounds are reflected by the dense concave bone of the cranium and an air sac at its base. The focused beam is modulated by a large fatty organ known as the 'melon'. This acts like an acoustic lens because it is composed of lipids of differing densities. Most toothed whales use clicks in a series, or click train, for echolocation, while the sperm whale may produce clicks individually. Toothed whale whistles do not appear to be used in echolocation. Different rates of click production in a click train give rise to the familiar barks, squeals and growls of the bottlenose dolphin. A click train with a repetition rate over 600 per second is called a burst pulse. In bottlenose dolphins, the auditory brain response resolves individual clicks up to 600 per second, but yields a graded response for higher repetition rates.

It has been suggested that some smaller toothed whales may have their tooth arrangement suited to aid in echolocation. The placement of teeth in the jaw of a bottlenose dolphin, as an example, are not symmetrical when seen from a vertical plane, and this asymmetry could possibly be an aid in the dolphin sensing if echoes from its biosonar are coming from one side or the other. However, this idea lacks experimental support.

Echoes are received using complex fatty structures around the lower jaw as the primary reception path, from where they are transmitted to the middle ear via a continuous fat body. Lateral sound may be received though fatty lobes surrounding the ears with a similar density to water. Some researchers believe that when they approach the object of interest, they protect themselves against the louder echo by quietening the emitted sound. In bats this is known to happen, but here the hearing sensitivity is also reduced close to a target.

Before the echolocation abilities of "porpoises" were officially discovered, Jacques Yves Cousteau suggested that they might exist. In his first book, *The Silent World* (1953, pp. 206–207), he reported that his research vessel, the *Élie Monier,* was heading to the Straits of Gibraltar and noticed a group of porpoises following them. Cousteau changed course a few degrees off the optimal course to the center of the strait, and the porpoises followed for a few minutes, then diverged toward mid-channel again. It was obvious that they knew where the optimal course lay, even if the humans didn't. Cousteau concluded that the cetaceans had something like sonar, which was a relatively new feature on submarines.

Oilbirds and Swiftlets

A Palawan swiftlet (*Aerodramus palawanensis*) flies in complete darkness inside the Puerto Princesa subterranean river cave.

Oilbirds and some species of swiftlet are known to use a relatively crude form of echolocation compared to that of bats and dolphins. These nocturnal birds emit calls while flying and use the calls to navigate through trees and caves where they live.

Shrews, Tenrecs and Rats

Terrestrial mammals other than bats known to echolocate include two genera (*Sorex* and *Blarina*) of shrews and the tenrecs of Madagascar. These include the wandering shrew (*Sorex vagrans*), the common or Eurasian shrew (*Sorex araneus*), and the short-tailed shrew (*Blarina brevicauda*). The nature of shrew sounds unlike those of bats are low amplitude, broadband, multi-harmonic and frequency modulated. They contain no 'echolocation clicks' with reverberations and would seem to be used for simple, close range spatial orientation. In contrast to bats, shrews use echolocation only to investigate their habitat rather than additionally to pinpoint food. There is evidence that blinded laboratory rats can use echolocation to navigate mazes.

Echolocation Jamming

Echolocation systems are susceptible to interference known as echolocation jamming or sonar jamming. Jamming occurs when non-target sounds interfere with target echoes. Jamming can be purposeful or inadvertent and can be caused by the echolocation system itself, other echolocating animals, prey, or humans. Echolocating animals have evolved to minimize jamming; however, echolocation avoidance behaviors are not always successful.

ELECTRIC FISH

Electric eels are fish capable of generating an electrical field.

Audio recording of resting electric organ discharge of *Brachyhypopomus bennetti*.

An **electric fish** is any fish that can generate electric fields. A fish that can generate electric fields is said to be *electrogenic* while a fish that has the ability to detect electric fields is said to be *electroreceptive*. Most electrogenic fish are also electroreceptive. Electric fish species can be found both in the ocean and in freshwater rivers of South America (Gymnotiformes) and Africa (Mormyridae). Many fish such as sharks, rays and catfishes can detect electric fields and are thus electroreceptive, but they are not classified as electric fish because they cannot generate electricity. Most common bony fish (teleosts), including most fish kept in aquaria or caught for food, are neither electrogenic nor electroreceptive.

Video of a complete electric organ discharge. The electric field potential is represented on a sagittal across the modelled fish. Hot tones represent positive potential values, while cold tones represent negative electric potentials. The black line indicates the points where the potentials are zero.

Electric fish produce their electrical fields from a specialized structure called an electric organ. This is made up of modified muscle or nerve cells, which became specialized for producing bioelectric fields stronger than those that normal nerves or muscles produce. Typically this organ is located in the tail of the electric fish. The electrical output of the organ is called the *electric organ discharge* (EOD).

Strongly Electric Fish

Strongly electric fish are fish with an EOD that is powerful enough to stun prey. Typical examples are the electric eel, the electric catfishes, and electric rays. The amplitude of the signal can range from 10 to 600 volts with a current of up to 1 ampere, according to the surroundings, for example different conductances of salt and fresh water. To maximize the power delivered to the surroundings, the impedances of the electric organ and the water must be matched:

- Strongly electric marine fish gives low voltage, high current electric discharges. In salt water, a small voltage can drive a large current limited by the internal resistance of the electric organ. Hence, the electric organ consists of many electrocytes in parallel. Freshwater fish have high voltage, low current discharges.
- In freshwater, the power is limited by the voltage needed to drive the current through the large resistance of the medium. Hence, these fish have numerous cells in series.

Weakly Electric Fish

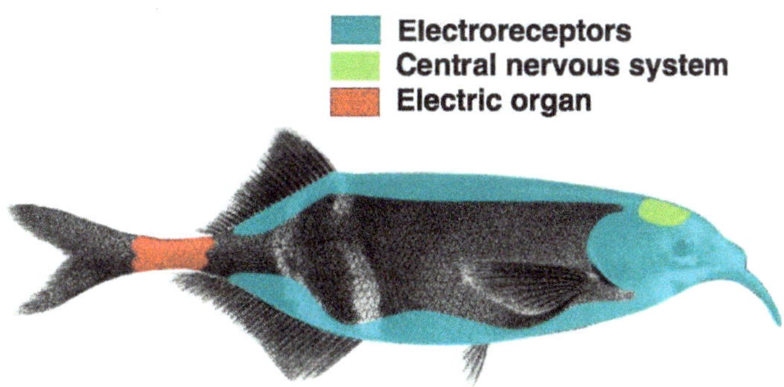

The elephantnose fish is a weakly electric fish which generates an electric field with its electric organ and then processes the returns from its electroreceptors to locate nearby objects.

Weakly electric fish generate a discharge that is typically less than one volt in amplitude. These are too weak to stun prey and instead are used for navigation, object detection (electrolocation) and communication with other electric fish (electrocommunication). Two of the best-known and most-studied examples are Peters' elephantnose fish (*Gnathonemus petersi*) and the black ghost knifefish (*Apteronotus albifrons*). The males of the nocturnal *Brachyhypopomus pinnicaudatus*, a toothless knifefish native to the Amazon basin, give off big, long electric hums to attract a mate.

The EOD waveform takes two general forms depending on the species. In some species the waveform is continuous and almost sinusoidal (for example the genera *Apteronotus*, *Eigenmannia* and *Gymnarchus*) and these are said to have a wave-type EOD. In other species, the EOD waveform consists of brief pulses separated by longer gaps (for example *Gnathonemus*, *Gymnotus*, *Leucoraja*) and these are said to have a pulse-type EOD.

Jamming Avoidance Response

It had been theorized as early as the 1950s that electric fish near each other might experience some type of interference or inability to segregate their own signal from those of neighbors. This issue does not arise, however, because the electric fish adjust to avoid frequency interference. In 1963, two scientists, Akira Watanabe and Kimihisa Takeda, discovered the behavior of the jamming avoidance response in the knifefish *Eigenmannia* sp. In collaboration with T.H. Bullock and colleagues, the behavior was further developed. Finally, the work of Walter Heiligenberg expanded it into a full neuroethology study by examining the series of neural connections that led to the behavior. *Eigenmannia* is a weakly electric fish that can self-generate electric discharges through electrocytes in its tail. Furthermore, it has the ability to electrolocate by analyzing the perturbations in its electric field. However when the frequency of a neighboring fish's current is very close (less than 20 Hz difference) to that of its own, the fish will avoid having their signals interfere through a behavior known as Jamming Avoidance Response. If the neighbor's frequency is higher than the fish's discharge frequency, the fish will lower its frequency, and vice versa. The sign of the frequency difference is determined by analyzing the "beat" pattern of the incoming interference which consists of the combination of the two fish's discharge patterns.

Neuroethologists performed several experiments under *Eigenmannia's* natural conditions to study how it determined the sign of the frequency difference. They manipulated the fish's discharge by injecting it with curare which prevented its natural electric organ from discharging. Then, an electrode was placed in its mouth and another was placed at the tip of its tail. Likewise, the neighboring fish's electric field was mimicked using another set of electrodes. This experiment allowed neuroethologists to manipulate different discharge frequencies and observe the fish's behavior. From the results, they were able to conclude that the electric field frequency, rather than an internal frequency measure, was used as a reference. This experiment is significant in that not only does it reveal a crucial neural mechanism underlying the behavior but also demonstrates the value neuroethologists place on studying animals in their natural habitats.

Species

Following is a table of electric fish species listed by family. Most families inhabit fresh water. Two groups of marine fish, the electric rays (Torpediniformes: Narcinidae and Torpedinidae) and the stargazers (Perciformes: Uranoscopidae), are capable of generating strong electric pulses.

BARN OWL

The **barn owl** (*Tyto alba*) is the most widely distributed species of owl and one of the most widespread of all birds. It is also referred to as the **common barn owl**, to distinguish it from other species in its family, Tytonidae, which forms one of the two main lineages of living owls, the other being the typical owls (Strigidae). The barn owl is found almost everywhere in the world except polar and desert regions, Asia north of the Himalayas, most of Indonesia, and some Pacific islands.

Phylogenetic evidence shows that there are at least three major lineages of barn owl, one in Eurasia and Africa, one in Australasia and one in the Americas, and some highly divergent taxa on islands. Some authorities further split the group, recognising up to five species, and further research needs to be done to clarify the position. There is a considerable variation between the sizes and colour of the approximately 28 subspecies but most are between 33 and 39 cm (13 and 15 in) in length with wingspans ranging from 80 to 95 cm (31 to 37 in). The plumage on head and back is a mottled shade of grey or brown, the underparts vary from white to brown and are sometimes speckled with dark markings. The face is characteristically heart-shaped and is white in most subspecies. This owl does not hoot, but utters an eerie, drawn-out shriek.

The barn owl is nocturnal over most of its range, but in Britain and some Pacific islands, it also hunts by day. Barn owls specialise in hunting animals on the ground and nearly all of their food consists of small mammals which they locate by sound, their hearing being very acute. They mate for life unless one of the pair gets killed, when a new pair bond may be formed. Breeding takes place at varying times of year according to locality, with a clutch, averaging about four eggs, being laid in a nest in a hollow tree, old building or fissure in a cliff. The female does all the incubation, and she and the young chicks are reliant on the male for food. When large numbers of small prey are readily available, barn owl populations can expand rapidly, and globally the bird is considered to be of least conservation concern. Some subspecies with restricted ranges are more threatened.

Taxonomy and Etymology

Barn owl in flight.

Male *Tyto alba alba* (left) and female *T. a. guttata* barn owls in the Netherlands, where these subspecies intergrade.

The barn owl was one of several species of bird first described in 1769 by the Tyrolean physician and naturalist Giovanni Antonio Scopoli in his *Anni Historico-Naturales*. He gave it the scientific name *Strix alba*. As more species of owl were described, the genus name *Strix* came to be used solely for the wood owls in the typical owl family Strigidae, and the barn owl became *Tyto alba* in the barn owl family Tytonidae. The name literally meanswhite owl, from the onomatopoeic Ancient Greek *tyto* (τυτώ) for an owl – compare Englishhooter – and Latin *alba*, "white". The bird is known by many common names which refer to its appearance, call, habitat, or its eerie, silent flight: white owl, silver owl, demon owl, ghost owl, death owl, night owl, rat owl, church owl, cave owl, stone owl, monkey-faced owl, hissing owl, hobgoblin or hobby owl, dobby owl, white-breasted owl, golden owl, scritch owl, screech owl, straw owl, barnyard owl, and delicate owl. "Golden owl" might also refer to the related golden masked owl (*T. aurantia*). "Hissing owl" and, particularly in the UK and in India, "screech owl", refers to the piercing calls of these birds. The latter name is also applied to a different group of birds, the screech-owls in the genus *Megascops*.

The ashy-faced owl (*T. glaucops*) was for some time included in *T. alba*, and by some authors its populations from the Lesser Antilles still are. Based on DNA evidence, König, Weick & Becking (2009) recognised the American Barn Owl (*T. furcata*) and the Curaçao Barn Owl (*T. bargei*) as separate species. They also proposed that *T. a. delicatula* should be split off as a separate species, to be known as the eastern barn owl, which would include the subspecies *T. d. sumbaensis*, *T. d. meeki*, *T. d. crassirostris* and *T. d. interposita*. However, the International Ornithological Committee has doubts about this and states that the split of *Tyto delicatula* from *T. alba* "may need to be revisited". Some island subspecies are occasionally treated as distinct species, a move which should await further research into barn owl phylogeography. According to Bruce in the *Handbook of Birds of the World, Volume 5: Barn-owls to Hummingbirds*, "a review of the whole group long overdue". Molecular analysis of mitochondrial DNA shows a separation of the species into two clades, an Old World *alba* and a New World *furcata*, but this study did not include *T. a. delicatula* which the authors seem to have accepted as a separate species. A high amount of genetic variation was also found between the Indonesian *T. a. stertens* and other members of the *alba* clade.

The barn owl has a wider distribution than any other species of owl. Many subspecies have been proposed over the years but several are generally considered to be intergrades between more distinct populations. Twenty to thirty are usually recognised, varying mainly in body proportions, size and colour. Island forms are mostly smaller than mainland ones, and those inhabiting forests have darker plumage and shorter wings than those occurring in open grasslands. Barn owls range in colour from the almost beige-and-white nominate subspecies, *erlangeri* and *niveicauda*, to the nearly black-and-brown *contempta*.

In the *Handbook of Birds of the World, Volume 5: Barn-owls to Hummingbirds*, the following subspecies are listed:

Description

The barn owl is a medium-sized, pale-coloured owl with long wings and a short, squarish tail. There is considerable size variation across the subspecies with a typical specimen measuring about 33 to 39 cm (13 to 15 in) in overall length, with a full range of 29 to 44 cm (11 to 17 in) across the species. Barn owls have a typical wingspan of some 80 to 95 cm (31 to 37 in), with a full range of 68 to 105 cm (27 to 41 in). Adult body mass is also variable with male owls from the Galapagos (*T. a. punctatissima*) weighing approximately 260 g (9.2 oz) on average while male Pacific barn owls (*T. a. javanica*) average 555 g (19.6 oz). The full known weight range for the barn owl species can range from 224 to 710 g (7.9 to 25.0 oz). In general, owls living on small islands are smaller and lighter, perhaps because they have a higher dependence on insect prey and need to be more manoeuvrable. However, the largest bodied race of barn owl, *T. a. furcata* from Cuba and Jamaica, is also an island race, albeit being found on more sizeable islands with larger prey and few larger owls competing for dietary resources.The shape of the tail is a means of distinguishing the barn owl from typical owls when seen in the air. Other distinguishing features are the undulating flight pattern and the dangling, feathered legs. The pale face with its heart shape and black eyes give the flying bird a distinctive appearance, like a flat mask with oversized, oblique black eye-slits, the ridge of feathers above the beak somewhat resembling a nose.

The bird's head and upper body typically vary between pale brown and some shade of grey (especially on the forehead and back) in most subspecies. Some are purer, richer brown instead, and all have fine black-and-white speckles except on the remiges and rectrices (main wing feathers), which are light brown with darker bands. The heart-shaped face is usually bright white, but in some subspecies it is brown. The underparts, including the tarsometatarsal (lower leg) feathers, vary from white to reddish buff among the subspecies, and are either mostly unpatterned or bear a varying number of tiny blackish-brown speckles. It has been found that at least in the continental European populations, females with more spotting are healthier than plainer birds. This does not hold true for European males by contrast, where the spotting varies according to subspecies. The beak varies from pale horn to dark buff, corresponding to the general plumage hue, and the iris is blackish brown. The talons, like the beak, vary in colour, ranging from pink to dark pinkish-grey and the talons are black.

On average within any one population, males tend to have fewer spots on the underside and are paler in colour than females. The latter are also larger with a strong female *T. alba* of a large subspecies weighing over 550 g (19.4 oz), while males are typically about 10% lighter. Nestlings are covered in white down, but the heart-shaped facial disk becomes visible soon after hatching.

Contrary to popular belief, the barn owl does not hoot (such calls are made by typical owls, like the tawny owl or other members of the genus *Strix*). It instead produces the characteristic *shree* scream, ear-shattering at close range, an eerie, long-drawn-out shriek. Males in courtship give a shrill twitter. Both young and old can hiss like a snake to scare away intruders. Other sounds produced include a purring chirrup denoting pleasure, and a "kee-yak, which resembles one of the vocalisations of the tawny owl. When captured or cornered, the barn owl throws itself on its back and flails with sharp-taloned feet, making for an effective defence. In such situations it may emit rasping sounds or clicking snaps, produced probably by the beak but possibly by the tongue.

Distribution

The barn owl is the most widespread landbird species in the world, occurring in every continent except Antarctica. Its range includes all of Europe (except Fennoscandia and Malta), most of Africa apart from the Sahara, the Indian subcontinent, Southeast Asia,

Australia, many Pacific Islands, and North, Central and South America. In general it is considered to be sedentary, and indeed many individuals, having taken up residence in a particular location, remain there even when better foraging areas nearby become vacant. In the British Isles, the young seem largely to disperse along river corridors and the distance travelled from their natal site averages about 9 km (5.6 mi).

In continental Europe the distance travelled is greater, commonly somewhere between 50 and 100 kilometres (31 and 62 mi) but exceptionally 1,500 km (932 mi), with ringed birds from the Netherlands ending up in Spain and in Ukraine. In the United States, dispersal is typically over distances of 80 and 320 km (50 and 199 mi), with the most travelled individuals ending up some 1,760 km (1,094 mi) from the point of origin. Movements in the African continent include 1,000 km (621 mi) from Senegambia to Sierra Leone and up to 579 km (360 mi) within South Africa. In Australia there is some migration as the birds move towards the northern coast in the dry season and southward in the wet, and also nomadic movements in association with rodent plagues. Occasionally, some of these birds turn up on Norfolk Island, Lord Howe Island or New Zealand, showing that crossing the ocean is not beyond their capabilities. In 2008, barn owls were recorded for the first time breeding in New Zealand. The barn owl has been successfully introduced into the Hawaiian island of Kauai in an attempt to control rodents, however it has been found to also feed on native birds.

Behaviour and Ecology

A barn owl's talon

Like most owls, the barn owl is nocturnal, relying on its acute sense of hearing when hunting in complete darkness. It often becomes active shortly before dusk and can sometimes be seen during the day when relocating from one roosting site to another. In

Britain, on various Pacific Islands and perhaps elsewhere, it sometimes hunts by day. This practice may depend on whether the owl is mobbed by other birds if it emerges in daylight. However, in Britain, some birds continue to hunt by day even when mobbed by such birds as magpies, rooks and black-headed gulls, such diurnal activity possibly occurring when the previous night has been wet making hunting difficult. By contrast, in southern Europe and the tropics, the birds seem to be almost exclusively nocturnal, with the few birds that hunt by day being severely mobbed.

Barn owls are not particularly territorial but have a home range inside which they forage. For males in Scotland this has a radius of about 1 km (0.6 mi) from the nest site and an average size of about 300 hectares. Female home ranges largely coincide with that of their mates. Outside the breeding season, males and females usually roost separately, each one having about three favoured sites in which to conceal themselves by day, and which are also visited for short periods during the night. Roosting sites include holes in trees, fissures in cliffs, disused buildings, chimneys and haysheds and are often small in comparison to nesting sites. As the breeding season approaches, the birds move back to the vicinity of the chosen nest to roost.

Being mobbed by Brewer's blackbird in California

The barn owl is a bird of open country such as farmland or grassland with some interspersed woodland, usually at altitudes below 2,000 metres (6,600 ft) but occasionally as high as 3,000 metres (9,800 ft) in the tropics. This owl prefers to hunt along the edges of woods or in rough grass strips adjoining pasture. It has an effortless wavering flight as it quarters the ground, alert to the sounds made by potential prey. Like most owls, the barn owl flies silently; tiny serrations on the leading edges of its flight feathers and a hairlike fringe to the trailing edges help to break up the flow of air over the wings, thereby reducing turbulence and the noise that accompanies it. Hairlike extensions to the barbules of its feathers, which give the plumage a soft feel, also minimise noise produced during wingbeats. The behaviour and ecological preferences may differ slightly even among neighbouring subspecies, as shown in the case of the European *T. a. guttata* and

T. a. alba that probably evolved, respectively, in allopatric glacial refugia in southeastern Europe, and in Iberia or southern France.

Diet and Feeding

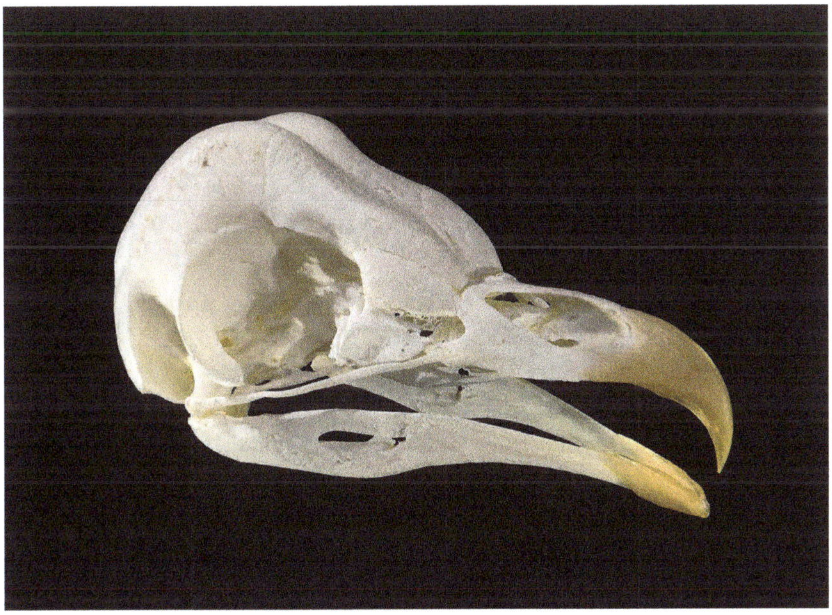

Skull, showing the powerful beak.

The diet of the barn owl has been much studied; the items consumed can be ascertained from identifying the prey fragments in the pellets of indigestible matter that the bird regurgitates. Studies of diet have been made in most parts of the bird's range, and in moist temperate areas over 90% of the prey tends to be small mammals, whereas in hot, dry, unproductive areas, the proportion is lower, and a great variety of other creatures are eaten depending on local abundance. Most prey is terrestrial but bats and birds are also taken, as well as lizards, amphibians and insects. Even when they are plentiful and other prey scarce, earthworms do not seem to be consumed.

In North America and most of Europe, voles predominate in the diet and shrews are the second most common food choice. Mice and rats form the main foodstuffs in the Mediterranean region, the tropics, sub-tropics and Australia. Barn owls are usually more specialist feeders in productive areas and generalists in drier areas. On the Cape Verde Islands, geckos are the mainstay of the diet, supplemented by birds such as plovers, godwits, turnstones, weavers and pratincoles, and on a rocky islet off the coast of California, a clutch of four young were being reared on a diet of Leach's storm petrel (*Oceanodroma leucorhoa*). In Ireland, the accidental introduction of the bank vole in the 1950s led to a major shift in the barn owl's diet: where their ranges overlap, the vole is now by far the largest prey item. Locally superabundant rodent species in the weight class of several grams per individual usually make up the single largest proportion of prey. In the United States, rodents and other small mammals usually make up ninety-five percent of the diet and worldwide, over ninety percent of the prey caught.

The barn owl hunts by flying slowly, quartering the ground and hovering over spots that may conceal prey. It may also use branches, fence posts or other lookouts to scan its surroundings, and this is the main means of prey location in the oil palm plantations of Malaysia. The bird has long, broad wings, enabling it to manoeuvre and turn abruptly. Its legs and toes are long and slender which improves its ability to forage among dense foliage or beneath the snow and gives it a wide spread of talons when attacking prey. Studies have shown that an individual barn owl may eat one or more voles (or their equivalent) per night, equivalent to about twenty-three percent of the bird's bodyweight. Excess food is often cached at roosting sites and can be used when food is scarce.

Small prey is usually torn into chunks and eaten completely including bones and fur, while prey larger than about 100 g (4 oz), such as baby rabbits, *Cryptomys*blesmols, or *Otomys* vlei rats, is usually dismembered and the inedible parts discarded. Contrary to what is sometimes assumed, the barn owl does not eat domestic animals on any sort of regular basis. Regionally, non-rodent foods are used as per availability. On bird-rich islands, a barn owl might include some fifteen to twenty percent of birds in its diet, while in grassland it will gorge itself on swarming termites, or on Orthoptera such as Copiphorinae katydids, Jerusalem crickets (Stenopelmatidae) or true crickets (Gryllidae). Bats and even frogs, lizards and snakes may make a minor but significant contribution to the diet; small Soricomorpha like *Suncus* shrews may be a secondary prey of major importance.

The barn owl has acute hearing, with ears placed asymmetrically. This improves detection of sound position and distance and the bird does not require sight to hunt. The facial disc plays a part in this process, as is shown by the fact that with the ruff feathers removed, the bird can still locate the source in azimuth but fails to do so in elevation. Hunting nocturnally or crepuscularly, this bird can target its prey and dive to the ground, penetrating its talons through snow, grass or brush to seize small creatures with deadly accuracy. Compared to other owls of similar size, the barn owl has a much higher metabolic rate, requiring relatively more food. Weight for weight, barn owls consume more rodents—often regarded as pests by humans—than possibly any other creature. This makes the barn owl one of the most economically valuable wildlife animals for agriculture. Farmers often find these owls more effective than poison in keeping down rodent pests, and they can encourage barn owl habitation by providing nest sites.

Breeding

Barn owls living in tropical regions can breed at any time of year, but some seasonality in nesting is still evident. Where there are distinct wet and dry seasons, egg-laying usually takes place during the dry season, with increased rodent prey becoming available to the birds as the vegetation dies off. In arid regions, such as parts of Australia, breeding may be irregular and may happen in wet periods, triggered by temporary increases in the populations of small mammals. In temperate climates, nesting seasons become more distinct and there are some seasons of the year when no egg-laying takes place.

In Europe and North America, most nesting takes place between March and June when temperatures are increasing. The actual dates of egg-laying vary by year and by location, being correlated with the amount of prey-rich foraging habitat around the nest site and often with the phase of the rodent abundance cycle. An increase in rodent populations will usually stimulate the local barn owls to begin nesting; thus, even in the cooler parts of its range, two broods are often raised in a good year.

Egg, Collection Museum Wiesbaden, Germany.

Females are ready to breed at ten to eleven months of age although males sometimes wait till the following year. Barn owls are usually monogamous, sticking to one partner for life unless one of the pair dies. During the non-breeding season they may roost separately, but as the breeding season approaches they return to their established nesting site, showing considerable site fidelity. In colder climates, in harsh weather and where winter food supplies may be scarce, they may roost in farm buildings and in barns between hay bales, but they then run the risk that their selected nesting hole may be taken over by some other, earlier-nesting species. Single males may establish feeding territories, patrolling the hunting areas, occasionally stopping to hover, and perching on lofty eminences where they screech to attract a mate. Where a female has lost her mate but maintained her breeding site, she usually seems to manage to attract a new spouse.

Once a pair-bond has been formed, the male will make short flights at dusk around the nesting and roosting sites and then longer circuits to establish a home range. When he is later joined by the female, there is much chasing, turning and twisting in flight, and frequent screeches, the male's being high-pitched and tremulous and the female's lower and harsher. At later stages of courtship, the male emerges at dusk, climbs high into the sky and then swoops back to the vicinity of the female at speed. He then sets off to forage. The female meanwhile sits in an eminent position and preens, returning to the nest a minute or two before the male arrives with food for her. Such feeding behaviour of the female by the male is common, helps build the pair-bond and increases the female's fitness before egg-laying commences..

An *Eulenloch* ("owl-hole) in northern Germany lets barn owls access the attic for nesting.

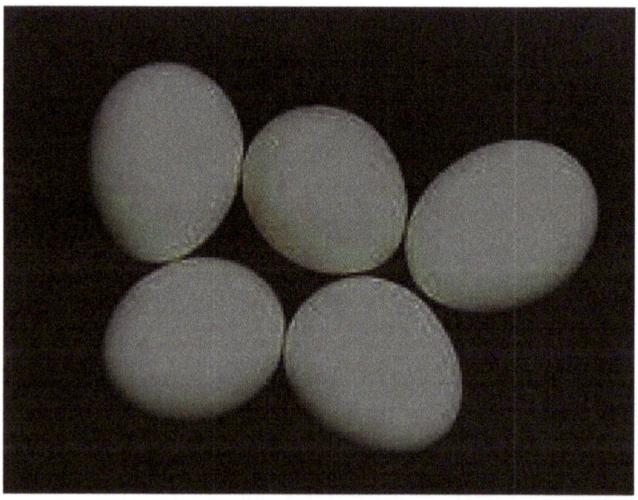

Clutch of eggs.

Barn owls are cavity nesters. They choose holes in trees, fissures in cliff faces, the large nests of other birds such as the hamerkop (*Scopus umbretta*) and, particularly in Europe and North America, old buildings such as farm sheds and church towers. Buildings are preferred to trees in wetter climates in the British Isles and provide better protection for fledglings from inclement weather. Trees tend to be in open habitats rather than in the middle of woodland and nest holes tend to be higher in North America than in Europe because of possible predation by raccoons (*Procyon lotor*). No nesting material is used as such but, as the female sits incubating the eggs, she draws in the dry furry material of which her regurgitated pellets are composed, so that by the time the chicks are hatched, they are surrounded by a carpet of shredded pellets. Oftentimes other birds such as jackdaws (*Corvus monedula*) nest in the same hollow tree or building and seem to live harmoniously with the owls.

Before commencing laying, the female spends much time near the nest and is entirely provisioned by the male. Meanwhile, the male roosts nearby and may cache any prey that is surplus to their requirements. When the female has reached peak weight, the male provides a ritual presentation of food and copulation occurs at the nest. The female lays eggs on alternate days and the clutch size averages about five eggs (range two to nine). The eggs are chalky white, somewhat elliptical and about the size of bantam's eggs, and incubation begins as soon as the first egg is laid. While she is sitting on the nest, the male is constantly bringing more provisions and they may pile up beside the female. The incubation period is about thirty days, hatching takes place over a prolonged period and the youngest chick may be several weeks younger than its oldest sibling. In years with plentiful supplies of food, there may be a hatching success rate of about 75%. The male continues to copulate with the female when he brings food which makes the newly hatched chicks vulnerable to injury.

Brood prior to fledging, beginning to shed their nestling down.

The chicks are at first covered with greyish-white down and develop rapidly. Within a week they can hold their heads up and shuffle around in the nest. The female tears up the food brought by the male and distributes it to the chicks. Initially these make a "chittering" sound but this soon changes into a food-demanding "snore". By two weeks old they are already half their adult weight and look naked as the amount of down is insufficient to cover their growing bodies. By three weeks old, quills are starting to push through the skin and the chicks stand, making snoring noises with wings raised and tail stumps waggling, begging for food items which are now given whole. The male is the main provider of food until all the chicks are at least four weeks old at which time the female begins to leave the nest and starts to roost elsewhere. By the sixth week the chicks are as big as the adults but have slimmed down somewhat by the ninth week when they are fully fledged and start leaving the nest briefly themselves. They are still dependent on the parent birds until about thirteen weeks and receive training from the female in finding, and eventually catching, prey.

Moulting

Feathers become abraded over time and all birds need to replace them at intervals. Barn owls are particularly dependent on their ability to fly quietly and manoeuvre efficiently, and in temperate areas their prolonged moult lasts through three phases over a period of two years. The female starts to moult while incubating the eggs and brooding the chicks, a time when the male feeds her so she does not need to fly much. The first primary feather to be shed is the central one, number 6, and it has regrown completely by the time the female resumes hunting. Feathers 4, 5, 7 and 8 are dropped at a similar time the following year and feathers 1, 2, 3, 9 and 10 in the bird's third year of adulthood. The secondary and tail feathers are lost and replaced over a similar timescale, again starting while incubation is taking place. In the case of the tail, the two outermost tail feathers are first shed followed by the two central ones, the other tail feathers being moulted the following year.

In temperate areas, the male owl moults rather later in the year than the female, at a time when there is an abundance of food, the female has recommenced hunting and the demands of the chicks are lessening. Unmated males without family responsibilities often start losing feathers earlier in the year. The moult follows a similar prolonged pattern to that of the female and the first sign that the male is moulting is often when a tail feather has been dropped at the roost. A consequence of moulting is the loss of thermal insulation. This is of little importance in the tropics and barn owls here usually moult a complete complement of flight feathers annually. The hot-climate moult may still take place over a long period but is usually concentrated at a particular time of year outside the breeding season.

Predators and Parasites

Predators of the barn owl include large American opossums (*Didelphis*), the common raccoon, and similar carnivorous mammals, as well as eagles, larger hawks and other owls. Among the latter, the great horned owl (*Bubo virginianus*) in the Americas and the Eurasian eagle-owl (*B. bubo*) are noted predators of barn owls. Despite some sources claiming that there is little evidence of predation by great horned owls, one study from Washington found that 10.9% of the local great horned owl's diet was made up of barn owls. In Africa, the principle predators of barn owls are Verreaux's eagle-owls and cape eagle-owls. In Europe, although less dangerous than the eagle owl, the chief diurnal predators are the northern goshawk (*Accipiter gentilis*) and the common buzzard (*Buteo buteo*). About a dozen other large diurnal raptors and owls have also been reported as predators of barn owls, ranging from the scarcely larger tawny owl to the golden eagle. The goshawk and the eagle owl are on the increase because of the greater protection these birds now receive.

Three barn owl chicks threatening an intruder.

When disturbed at its roosting site, an angry barn owl lowers its head and sways it from side to side, or the head may be lowered and stretched forward and the wings drooped while the bird emits hisses and makes snapping noises with its beak. A defensive attitude involves lying flat on the ground or crouching with wings spread out.

Barn owls are hosts to a wide range of parasites. Fleas are present at nesting sites and externally the birds are attacked by feather lice and feather mites which chew the barbules of the feathers and which are transferred from bird to bird by direct contact. Blood-sucking flies such as *Ornithomyia avicularia* are often present, moving about among the plumage. Internal parasites include the fluke *Strigea strigis*, the tape worm *Paruternia candelabraria*, several species of parasitic round worm and spiny-headed worms in the genus *Centrorhynchus*. These gut parasites are acquired when the birds feed on infected prey which provide intermediate hosts for the parasites. There is some indication that female birds with more and larger spots have a greater resistance to external parasites. This is correlated with smaller bursae of Fabricius, glands associated with antibody production, and a lower fecundity of the blood-sucking fly *Carnus hemapterus* that attacks nestlings.

Lifespan

Landing on a handler's gloved hand. Captive birds often live longer than wild ones.

Unusually for such a medium-sized carnivorous animal, the barn owl exhibits *r*-selection, producing large number of offspring with a high growth rate, many of which have a relatively low probability of surviving to adulthood. While wild barn owls are thus decidedly short-lived, the actual longevity of the species is much higher – captive individuals may reach twenty years of age or more. But occasionally, a wild bird reaches an advanced age. The American record age for a wild barn owl is eleven and a half years, while a Dutch bird was noted to have reached an age of seventeen years, ten months. Another captive barn owl, in England, lived to be over twenty-five years old. Taking into account such extremely long-lived individuals, the average lifespan of the barn owl is about four years, and statistically two-thirds to three-quarters of all adults survive from one year to the next. However, the mortality is not evenly distributed throughout the bird's life, and only one young in three manages to live to its first breeding attempt.

The most significant cause of death in temperate areas is likely to be starvation, particularly over the autumn and winter period when first year birds are still perfecting their hunting skills. In northern and upland areas, there is some correlation between mortality in older birds and adverse weather, deep-lying snow and prolonged low temperatures. Collision with road vehicles is another cause of mortality, and may result when birds forage on mown verges. Some of these birds are in poor condition and may have been less able to evade oncoming vehicles than fit individuals would have been. Historically, many deaths were caused by the use of pesticides, and this may still be the case in some parts of the world. Collisions with power-lines kill some birds and shooting accounts for others, especially in Mediterranean regions.

Status and Conservation

Barn owls are relatively common throughout most of their range and not considered globally threatened. However, locally severe declines from organochlorine (e.g., DDT) poisoning in the mid-20th century and rodenticides in the late 20th century have affected some populations, particularly in Europe and North America. Intensification of agricultural practices often means that the rough grassland that provides the best foraging habitat is lost. While barn owls are prolific breeders and able to recover from short-term population decreases, they are not as common in some areas as they used to be. A 1995–1997 survey put their British population at between 3,000 and 5,000 breeding pairs, out of an average of about 150,000 pairs in the whole of Europe. In the USA, barn owls are listed as endangered species in seven Midwestern states, and in the European Community they are considered a Species of European Concern.

Barn owl on Lithuanian silver coin of 5 litas (2002).

In some areas, it may be an insufficiency of suitable nesting sites that is the factor limiting barn owl numbers. The provision of nest boxes under the eaves of buildings and in other locations can be very successful in increasing the local population. In Malaysia, large areas of rainforest were felled to make way for oil palm plantations and with few tree cavities for breeding, the barn owl population, with its ability to control rodent pests, diminished. The provision of two hundred nest boxes in a trial saw almost one hundred percent occupancy and as the programme expanded, the plantations supported one of the densest barn owl populations in the world. Similarly, providing nesting boxes has increased the number of barn owls in rice-growing areas of Malaysia where the rodents do much damage to the crop. However, although barn owl numbers have increased in both these instances, it is unclear as to how effective this biological control of the rats is as compared to the trapping and baiting that occurred previously.

Common names such asdemon owl,death owl, orghost owl show that traditionally, rural populations in many places considered barn owls to be birds of evil omen. Consequently, they were often persecuted by farmers who were unaware of the benefit these birds bring. The Canary barn owl is particularly at risk, and as late as 1975, hunting by fearful locals was limiting the population on Fuerteventura where only a few dozen pairs remain. On Lanzarote a somewhat larger number of these birds still seem to exist, but altogether this particular subspecies is precariously rare: Probably less than three hundred and perhaps fewer than two hundred individuals still remain. Similarly, the birds on the western Canary Islands which are usually assigned to the nominate subspecies have declined much, and here wanton destruction seems still to be significant. On Tenerife they seem relatively numerous but on the other islands, the situation looks about as bleak as on Fuerteventura. Due to their assignment to the nominate subspecies, which is common in mainland Spain, the western Canary Islands population is not classified as threatened.

In the United Kingdom, theBarn Owl Nest Box Scheme is promoted by the World Owl Trust and has many participants in local areas such as Somerset, where a webcam has been set up inside a nest box in which seven young were reared in 2014. Another barn owl nest box live-streaming webcam located in California, United States has proved popular online. In May 2012, it was revealed that farmers in Israel and Jordan had, over a period of ten years, replaced pesticides with barn owls in a joint conservation venture called "Project Barn Owl".

VISION IN TOADS

The neural basis of prey detection, recognition, and orientation was studied in depth by Jörg-Peter Ewert in a series of experiments that made the toad visual system a model system in neuroethology (neural basis of natural behavior). He began by observing the natural prey catching behavior of the common European toad (*Bufo bufo*).

Ewert's work with toads yielded several important discoveries (Ewert 1974, 2004). In general, his research revealed the specific neural circuits for recognition of complex visual stimuli. Specifically, he identified two main regions of the brain, the tectum and the thalamic-pretectal region, that were responsible for discriminating prey from non-prey and revealed the neural pathways that connected them. Furthermore, he found that the neural mechanisms are plastic and adaptable to varying environments and conditions (Carew 2000; Zupanc 2004).

The eye of a toad.

Natural Toad Behavior

Bufo bufo.

The common toad responds to a moving insect or worm with a series of prey-catching reactions: (1) orienting towards prey, (2) stalking up to prey, (3) binocular fixation, (4) snapping, (5) swallowing and (6) mouth-wiping with forelimb (Ewert 1974). This series of movement constitutes a stimulus-response chain, where each reaction of the toad provides the stimulus constellation for the next response. First, if an object is recognized as prey and thus catches the toad's attention, the toad will orient towards the stimulus by turning its body to face it. Then it approaches the prey, binocularly focusing intently on it. During the attack, it snaps at the object with its tongue or jaws and swallows it. Finally, it wipes its mouth with a forelimb. These actions constitute a series of well-defined behavioral patterns.

One reason for this type of stimulus-response chain is that, unlike humans, toads do not have involuntary saccadic eye movements and they also cannot perform "tracking eye movements" (Ewert 1980). They must, therefore, depend on recognizing the stimulus before they respond. As a result, they have developed a specific detection system that, for example, allows them to discriminate between edible prey and dangerous predators.

The lack of saccadic eye movements forces the toad to hold its eyes in rigid positions. Therefore, it must decide whether the object is "prey" or "non-prey" before moving itself. If it orients towards an object, it must already have decided "prey" and then commits itself to snapping by reducing the thresholds for subsequent prey-catching responses. Even when the prey stimulus quickly disappeared after orienting, the aroused toad may sometimes complete the subsequent responses.

Prey vs. Predator Response

When a toad is presented with a moving stimulus, it generally may react with one of two responses. Depending on the size and the configuration of the stimulus, it will either engage in orienting (prey-catching) behavior or avoidance (escape) behavior, which consists of "planting-down" defensive postures or a crouching avoidance response.

In determining the size of a stimulus, a toad will consider both the angular size, which is measured in degrees of visual angle, and the absolute size, which takes into consideration the distance between the toad and the object. This second ability, to judge absolute size by estimating distance, is known as size constancy.

To study behavioral responses of toads to varying types of stimuli, Ewert conducted experiments by placing the toad in the center of a small cylindrical glass vessel. He then rotated a small stripe (bar) of contrasting cardboard (acting as a visual 'dummy') around the vessel to mimic either prey-like or threat-like stimuli; see Video. The rate of turning was recorded as a measure of orienting behavior (prey-catching activity). By changing characteristics of the visual stimulus in a methodical manner, Ewert was able to comprehensively study the key features that determine behavior.

Up to a certain size, squares rotated around the toad successfully elicited prey-catching responses. Toads avoided large squares. Vertical bars nearly never elicited prey-catching behavior and they were increasingly ineffective with increasing height. Horizontal bars, in contrast, were very successful at eliciting prey-catching behavior and their effectiveness increased with increasing length, to a certain degree. Additional vertical segments on top of horizontal bars significantly decreased prey-catching responses. In general, movement of a rectangle in the direction of its long axis is perceived by the toad to be wormlike, whereas movement along the short axis is interpreted as anti-wormlike. By means of a different experimental setup it was shown that the worm vs. anti-worm discrimination is independent (invariant) of the direction the object moves in the toad's visual field.

It is important to note that stationary objects usually elicit no prey-catching or avoidance responses. However, toads in an untextured surrounding cannot distinguish between the retinal image of a small moving object and the retinal image of a small stationary object induced by the toad's own motion. Both images elicit prey-capture. In a textured surrounding, however, the induced moving retinal image is ignored. — In addition, the contrast between stimuli and background can significantly affect the type of behavior. In response to a wormlike stripe, common toads orient and snap towards the edge leading in the direction of motion, given that the stripe is black and the background white. If the stimulus/background contrast is reversed, the toad prefers the trailing edge of the white stripe and often snaps behind it. Obviously, "off"-effects (rapid change in luminance from bright to dark) by the moving contrast borders play a guiding role. — Generally, white square objects moving against a black background are more attractive as prey than black objects on white. However, this tendency is plastic and reverses seasonally, where black objects against a white background are much more effective at eliciting prey-catching behavior in the fall and winter (Ewert 1980).

Feature Detectors and the Visual System

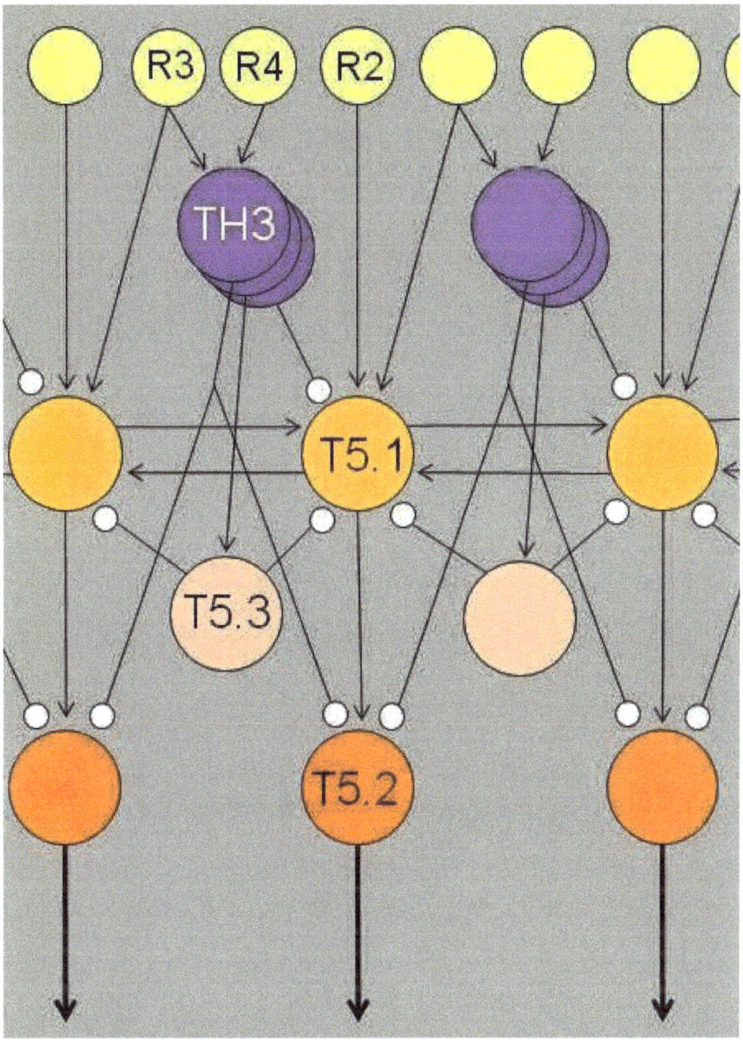

Tectal prey feature detectors T5.2 project the axons towards bulbar premotor/motor systems. Their response characteristic results from integration in a neuronal network involving retinal gangion cells R2, R3, R4, pretectal thalamic neurons TH3, and tectal neurons T5.1, T5.3. Arrows: excitatory connections; lines with terminal dots: inhibitory influences. Connections were checked, e.g., by means of intracellular recording postsynaptic potentials, horseradishperoxidase or cobalt-lysine tracing, antidromic electro-stimulation/recording, neurochemical techniques, chemical lesions by kainic acid. Hypothesis after Ewert 1974 and 2004.

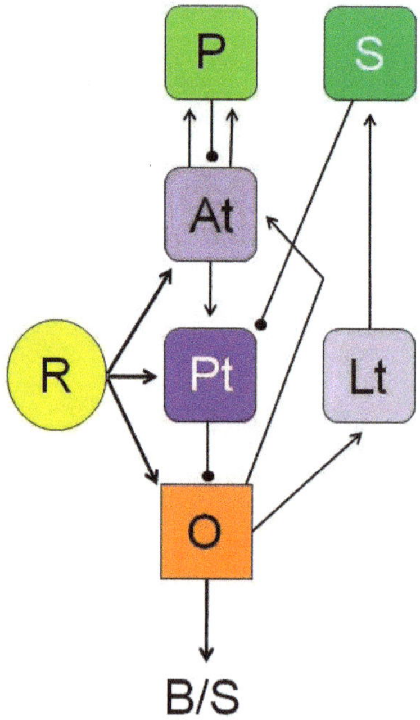

Visually released prey-catching in common toads: Various brain structures are involved in visual processes, such as feature discrimination (R, Pt, O), learning (P, At, Pt, O), and attention (S, Pt, Lt, O). The underlying neural network is suggested from experiments applying single neuron recording, focal brain stimulation, 14C-2-deoxiglucose labeling, lesioning, and anatomical pathway tracing. - R, retina (contralateral eye); O, optic tectum; Pt, pretectal thalamus; Lt, lateral anterior thalamic nucleus; At, anterior thalamus; P, posterior ventromedial pallium; S, caudal ventral striatum; B/S, bulbar/spinal pre-motor/motor nuclei; arrows: excitatory input; lines with dots: inhibitory influences. After Ewert and Schwippert 2006.

To understand the neural mechanisms underlying the toad's behavioral responses, Ewert performed a series of recording and stimulation experiments. First and foremost, the results allowed him to understand the way the visual system is constructed and connected to the central nervous system. Secondly, he discovered areas of the brain that were responsible for differential analysis of stimuli.

First, the retina is connected to the optic tectum by at least three types of ganglion cells, each with an excitatory receptive field and a surrounding inhibitory receptive field, but they differ in the diameter of their central excitatory receptive fields. Diameters in Class II (R2) ganglion cells are approximately four degrees visual angle. Those in Class III (R3) cells are about eight degrees and Class IV (R4) ganglion cells range from twelve to fifteen degrees. As stimuli move across the toad's visual field, information is sent to the optic tectum in the toad's midbrain. The optic tectum exists as an ordered localization system, in the form of a topographical map. Each point on the map corresponds to a particular region of the toad's retina and thus its entire visual field. Likewise, when a spot on the tectum was electrically stimulated, the toad would turn toward a corresponding part of its visual field, providing further evidence of the direct spatial connections.

Among Ewert's many experimental goals was the identification of feature detectors, neurons that respond selectively to specific features of a sensory stimulus. Results showed that there were no "worm-detectors" or "enemy-detectors" at the level of the retina. Instead, he found that the optic tectum and the thalamic-pretectal region (in the diencephalon) play significant roles in the analysis and interpretation of visual stimuli (summarized in Ewert 1974, 2004; Ewert and Schwippert 2006).

Electrical stimulation experiments demonstrated that the tectum initiates orienting and snapping behaviors. It contains many different visually sensitive neurons, among these Type I and Type II neurons (later named T5.1 and T5.2, respectively). Type I neurons are activated when an object traversing the toad's visual field is extended in the direction of movement; Type II neurons, too, but they will fire less when the object is extended in a direction that is perpendicular to the direction of movement. Those T5.2 neurons display prey-selective properties; see prey feature detectors. The discharge patterns of these neurons – recorded in freely moving toads – "predict" prey-catching reactions, e.g. the tongue flip of snapping. Their axons project down to the bulbar/spinal motor systems, such as the hypoglossal nucleus which harbors the motor neurons of the tongue muscles. In combination with additional projection neurons, prey-selective cells contribute to the ability of the tectum to initiate orienting behavior and snapping, respectively.

The thalamic-pretectal region initiates avoidance behavior in the toad. More specifically, electrical triggering the thalamic-pretectal region initiates a variety of protective movements such as eyelid closing, ducking and turning away (Ewert 1974, 2004). Various types of neurons in this region are responsible for the avoidance behaviors and they are all sensitive to different types of stimuli. One type of neurons (TH3) is activated by large threatening objects, especially those ones that are extended perpendicularly to the direction of motion. Another type (TH6) is activated by a looming object moving toward the toad. Still other types (TH10) respond to large stationary obstacles, and there are also neurons responding to stimulation of the balance sensors in the toad's ear. Stimulation of (a combination of) such types of neurons would cause the toad to display different kinds of protective behaviors.

Lesioning experiments led to the discovery of pathways extending between the tectum and the thalamic-pretectal region. When the tectum was removed, orienting behavior disappeared. When the thalamic-pretectal region was removed, avoidance behavior was entirely absent while orienting behavior was enhanced even to predator stimuli. Furthermore, prey-selective properties were impaired both in prey-selective neurons and in prey-catching behavior (Zupanc 2004). Finally, when one half of the thalamic-pretectal region was removed, the disinhibition applied to the entire visual field of the opposite eye. These and other experiments suggest that pathways, involving axons of type TH3 cells, extend from the pretectal thalamus to the tectum, suitable to modulate tectal responses to visual stimuli and to determine prey-selective properties due to inhibitory influences.

Modulatory loops and Evolutionary Perspectives

Having analyzed neuronal processing streams in brain structures (pretectal, tectal, medullary) that mediate between visual stimuli and adequate behavioral responses in toads, Ewert and coworkers examined various neural loops that — in connection with certain forebrain structures (striatal, pallial, thalamic) — can initiate, modulate or modify stimulus-response mediation (Ewert and Schwippert 2006). For example, in the course of associative learning the toad's visual prey schema can be modified to include non-prey objects. After lesions to a telencephalic structure involved in learning — the posterior ventromedial pallium — this learning effect fails and prey recognition shows again its species-specific selectivity. The posterior part of the ventromedial pallium is homologous to the hippocampus of mammals which is also involved in learning processes. Both in anuran amphibians and mammals striatal efferents are, for example, involved in directed attention, i.e. gating an orienting response towards a sensory stimulus. The anuran striatum is homologous to a portion of the amniote basal ganglia.

From an evolutionary point of view it is important to note that the tetrapod vertebrates share a common pattern of homologous brainstem and forebrain structures (e.g., see Reiner et al. 1998; González et al. 1999; Kenigfest et al. 2002). Neuroethological, neuroanatomical, and neurochemical investigations suggest that the neural networks underlying essential functions — such as attention, orienting, approaching, avoidance, associative or non-associative learning, and basic motor skills — have, so to speak, a phylogenetic origin in homologous structures of the amphibian brain.

From a neural network approach, it is reasonable to ask whether the toad's ability to classify moving objects by special configuration cues — object's dimension parallel vs. perpendicular to the direction of motion — is unique in the animal kingdom. Developmental studies suggest that this detection principle is an adaptation in terrestrial amphibians to their biotope and thus addressed to objects that are moving on land. Actually, it was found that this specific detection ability fails to occur in aquatic frogs (*Rana esculenta*). In common toads (*Bufo bufo*) it was shown that this detection property matures during the metamorphosis of the aquatic tadpoles in the course of their transition to terrestrial life (independent of food experience). Interestingly, comparable detection principles are discovered in amphibious fish (*Periophthalmus koelreuteri*) and in insects (*Sphodromantis lineola*). In mammals, erect body postures, for example, may address a threat signal to a rival. This suggests that the configuration-algorithm responsible for the distinction between profitable (e.g., prey-like) vs. dangerous (e.g., threat-like) may be implemented by quite different neural networks. Studies in artificial neuronal nets support this presumption (for an example see Ewert 2004).

REFERENCES

1. BirdLife International (2012).Tyto alba. IUCN Red List of Threatened Species. IUCN. 2012: e.T22688504A38682217. doi:10.2305/IUCN.UK.2012-1.RLTS.T22688504A38682217.en. Retrieved 11 August 2016.
2. Bruce (1999) pp. 34–75
3. Peters, James Lee (1964). Check-list of Birds of the World. Volume IV. Harvard University Press. p. 77.
4. Scopoli, Giovanni Antonio (1769).Strix alba. Annus I Historico-Naturalis *(in Latin). C. G.* Hilscheri. pp. 21–22.
5. Barn Owl Tyto alba (Scopoli, 1769). British Trust for Ornithology. 2009.

Retrieved 2014-09-08.

6. Ali, S. (1993). The Book of Indian Birds. Bombay: Bombay Natural History Society. ISBN 0-19-563731-3.
7. König, Claus; Weick, Friedhelm; Becking, Jan-Hendrik (2009). Owls of the World. Bloomsbury Publishing. pp. 46–48. ISBN 978-1-4081-0884-0.
8. König, Claus; Weick, Friedhelm; Becking, Jan-Hendrik (2009). Owls of the World. Bloomsbury Publishing. p. 209. ISBN 978-1-4081-0884-0.
9. Owls. IOC World Bird List: Version 4.3. IOC. Retrieved 2014-09-18.
10. Alaie Kakhki, N.; Aliabadian, M. (2012).Mitochondrial DNA (CYTB) divergences in two distinct, Old World and New World barn owls. Iranian Journal of Animal Biosystematics. 8 *(1): 47–55. ISSN 1735-434X.*
11. Taylor (2004) p. 24
12. Mátics, Róbert; Hoffmann, Gyula (2002).Location of the transition zone of the barn owl subspecies Tyto alba alba *and* Tyto alba guttata *(*Strigiformes: Tytonidae) (PDF). Acta Zoologica Cracoviensia. 45 *(2): 245–250.*
13. Olson, Storrs L.; James, Helen F.; Meister, Charles A. (1981).Winter field notes and specimen weights of Cayman Island Birds (PDF). Bulletin of the British Ornithologists' Club. 101 *(3): 339–346.*
14. Traylor, Melvin A.; Parelius, Daniel (1967).A collection of birds from the Ivory Coast. Fieldiana Zoology. 51 *(7): 91–117.*
15. Krabbe, Niels; Flórez, Pablo; Suárez, Gustavo; Castaño, José; Arango, Juan David; Duque, Arley (2006).The birds of Páramo de Frontino, western Andes of Colombia (PDF). Ornitología Colombiana. 4: 39–50.
16. Owls of the World: A Photographic Guide by Mikkola, H. Firefly Books (2012), ISBN 9781770851368
17. CRC Handbook of Avian Body Masses, 2nd Edition by John B. Dunning Jr. (Editor). CRC Press (2008), ISBN 978-1-4200-6444-5.
18. Taylor (2004) pp. 19–22
19. Bruce (1999) pp 34–75, Svensson et al. (1999) pp. 212–213
20. Taylor (2004) p. 169
21. Witherby (1943) pp. 343–347
22. Shawyer (1994) p. 10
23. Shawyer (1994) p. 91
24. Hyde, N. H. S.; Matthews, K.; Thompson, M.; Gale, R. (2009).First record of barn owls (Tyto alba) breeding in the wild in New Zealand. Notornis. 56 *(4): 169–175.*
25. Denny, Jim (2006).Introduced birds: Barn owl. Birds of Kaua'i. Retrieved 2014-07-24.
26. Martin, Jeff (2013-07-08).The daylight activity of barn owls. BritishBirds. BritishBirds Rarities Committee. Retrieved 2014-07-19.

27. Taylor (2004) pp. 96–107
28. Taylor (2004) pp. 47–61
29. Ehrlich et al. (1994) pp. 250–254
30. Taylor (2004) pp. 29–46
31. Naurois, R. de (1982).Le statut de la Chouette effraie de l'archipel du Cape Verte Tyto alba detorta. Rivista italiana di Ornitologia *(in French).* 52 *(3–4): 154–166.*
32. Bonnot, Paul (1928).An outlaw Barn Owl (PDF). Condor. 30 *(5): 320.* doi:10.2307/1363231. JSTOR 1363231.
33. Kelleher, K. M.; Oliver, G. A.; Sleeman, D. P. (2010).The composition and sex of rodent prey taken by Barn Owls Tyto alba *at a roost in County Cork". Irish Birds.* 9: 35–40. ISSN 0332-0111.
34. Laudet, Frédéric; Denys, Christiane; Senegas, Frank (2002).Owls, multirejection and completeness of prey remains: implications for small mammal taphonomy (PDF). Acta Zoologica Cracoviensia. 45: 341–355.
35. Ingles, Chuck (1995).Summary of California studies analyzing the diet of barn owls. Sustainable Agriculture/Technical Reviews. 2: 14–16. Archived from the original on November 28, 2011.
36. Taylor (2004) p. 29
37. Lavariega, Mario C.; García-Meza, Josué; Martínez-Ayón, Yazmín del Mar; Camarillo-Chávez, David; Hernández-Velasco, Teresa; Briones-Salas, Miguel (2015).Análisis de las presas de la Lechuza de Campanario (Tytonidae) en Oaxaca Central, México. Neotropical Biology and Conservation (in Spanish). 11 *(1). doi:10.4013/nbc.2016.111.03. ISSN 2236-3777.*
38. Taylor (2004) pp. 91–95
39. Knudsen, Eric I.; Konishi, Masakazu (1979).Mechanisms of sound localization in the barn owl (Tyto alba). Journal of Comparative Physiology. 133 *(1): 13–21.* doi:10.1007/BF00663106.
40. Meyrom, Kobi; Motro, Yoav; Leshem, Yossi; Aviel, Shaul; Izhaki, Ido; Argyle, Francis; Charter, Motti (2009).Nest-box use by the barn owl Tyto alba in a biological pest control program in the Beit She'an Valley, Israel. Ardea. 97 *(4): 463–467. doi:10.5253/078.097.0410.*
41. Marti, Carl D.; Poole, Alan F.; Bevier, L. R. (2005): Barn Owl (Tyto alba) The Birds of North America Online (A. Poole, Ed.). Ithaca: Cornell Lab of Ornithology; Barn owl
42. Millsap, B. A. and P. A. Millsap. 1987. Burow nesting by common Barn Owls in north central Colorado. Condor no. 89:668-670.
43. Knight, R. L. and R. E. Jackman. 1984. Food-niche relationships between Great Horned Owls and Common Barn-Owls in eastern Washington. Auk 101:175-179.
44. Steyn, P. (1983). Birds of prey of southern Africa: Their identification and life histories. Croom Helm, Beckenham (UK). 1983.

45. Brown, L. H. (1965). Observations on Verreaux's Eagle Owl Bubo lacteus (Temminck) in Kenya. Journal of the East African Natural History Society, 25, 101-107.

46. Voous, K.H. 1988. Owls of the Northern Hemisphere. The MIT Press, 0262220350.

47. Bunn, D. S.; Warburton, A. B.; Wilson, R. D. S. (2010). The Barn Owl. Bloomsbury Publishing. p. 177. ISBN 978-1-4081-3961-5.

48. Roulin, Alexandre; Riols, Christian; Dijkstra, Cor; Ducrest, Anne-Lyse (2001). Female plumage spottiness signals parasite resistance in the barn owl (Tyto alba). Behavioral Ecology. 12 *(1): 103–110. doi:10.1093/oxfordjournals.beheco. a000371.*

49. Altwegg, Res; Schaub, Michael; Roulin, Alexandre (2007).Age-specific fitness components and their temporal variation in the barn owl". The American Naturalist. 169 *(1): 47–61.* doi:10.1086/510215.

50. Taylor (2004) pp. 203–215

51. Taylor (2004) pp. 242–261

52. Wood, B. J.; Fee, Chung Gait (2003).A critical review of the development of rat control in Malaysian agriculture since the 1960s. Crop Protection. 22 *(3): 445–461.* doi:10.1016/S0261-2194(02)00207-7.

53. Spence, Cindy (1999-10-28).Spooky owl provides natural rodent control for farmers. UF News. University of Florida. Archived from the original on 2014-03-07. Retrieved 2014-07-14.

54. *Álamo* Tavío, Manuel (1975).Aves de Fuerteventura en peligro de extinción. Aves y plantas de Fuerteventura en peligro de extinción *(in Spanish).* Asociación Canaria para Defensa de la Naturaleza. pp. 10–32.

55. Tyto alba gracilirostris Hartert, 1905 (PDF) (in Spanish). El Ministerio de Medio Ambiente agradece sus comentarios, Spain. 2006. Retrieved 2014-09-19.

56. Palacios, César-Javier (2004).Current status and distribution of birds of prey in the Canary Islands (PDF). Bird Conservation International. 14 *(3): 203–213.* doi:10.1017/S0959270904000255.

57. Nest-boxes for Barn Owls. World Owl Trust. Archived from the original on June 12, 2015. Retrieved 2014-07-24.

58. Barn Owl Project. Somerset Wildlife Trust. Retrieved 2014-07-24.

59. Owl Cam a Hoot Online. CBS News. 2010-04-12. Retrieved 2014-10-02.

60. Santorelli, Tom (2012-05-18). "The Israeli-Jordan barn owl love that knows no borders". BBC News. Retrieved 2014-07-24.

61. Surlykke, A; Kalko, EK (2008). "Echolocating bats cry out loud to detect their prey.". PLoS ONE. 3 *(4): e2036. doi:10.1371/journal.pone.0002036. PMC 2323577. PMID 18446226.*

62. Kick, S. A.; J. A. Simmons (1984).Automatic gain control in the bat's sonar receiver and the neuroethology of echolocation. Journal of Neuroscience. 4 *(11): 2725–37. PMID 6502201.*

63. Henson OW, Jr (1965).The activity and function of the middle-ear muscles in echo-locating bats.. *The Journal of Physiology.* 180 *(4): 871–887. doi:10.1113/jphysiol.1965.sp007737. PMID 5880367.*

64. Griffin, Donald R.; Webster, Frederic A.; Michael, Charles R. (1960).The echolocation of flying insects by bats. Animal Behaviour. 8 *(3-4): 141–154.* doi:10.1016/0003-3472(60)90022-1.

65. Thomas, Jeanette A.; Moss, Cynthia F.; Vater, Marianne, eds. (2004). Echolocation in bats and dolphins. Chicago: The University of Chicago Press. pp. 22–26. ISBN 0226795985.

66. Elemens, Coen P. H.; A. F. Mead; L. Jakobsen; J. M. Ratcliffe (2011).Superfast Muscles Set Maximum Call Rate in Echolocating Bats. Science. 333 *(6051): 1885–1888. doi:10.1126/science.1207309. PMID 21960635.*

67. Hiryu, S.; Bates, M. E.; Simmons, J. A.; Riquimaroux, H. (2010).FM echolocating bats shift frequencies to avoid broadcast-echo ambiguity in clutter. Proceedings of the National Academy of Sciences. 107 *(15): 7048–7053. doi:10.1073/pnas.1000429107.*

68. Ulanovsky, N.; Fenton, M. B.; Tsoar, A.; Korine, C. (2004).Dynamics of jamming avoidance in echolocating bats. Proceedings of the Royal Society B: Biological Sciences. 271 *(1547): 1467–1475. doi:10.1098/rspb.2004.2750.*

69. Gillam, E. H; Ulanovsky, N.; McCracken, G. F (2007).Rapid jamming avoidance in biosonar. Proceedings of the Royal Society B: Biological Sciences. 274 *(1610): 651–660. doi:10.1098/rspb.2006.0047.*

70. Chiu, C.; Xian, W.; Moss, C. F. (2008).Flying in silence: Echolocating bats cease vocalizing to avoid sonar jamming. Proceedings of the National Academy of Sciences. 105 *(35): 13116–13121. doi:10.1073/pnas.0804408105.*

71. Barber, J. R. (2006).Tiger moth responses to a simulated bat attack: timing and duty cycle. Journal of Experimental Biology. 209 *(14): 2637–2650. doi:10.1242/jeb.02295.*

72. Hristov, N. I; W. E. Conner (2005).Sound strategy: acoustic aposematism in the bat–tiger moth arms race. Naturwissenschaften. 92 *(4): 164–169.* doi:10.1007/s00114-005-0611-7. PMID 15772807.

73. Corcoran, Aaron J.; J. R. Barber; W. E. Conner (2009).Tiger moth jams bat sonar. Science. 325 *(5938): 325–327. doi:10.1126/science.1174096.*

74. Corcoran, Aaron J.; W. E. Conner (2012).Sonar jamming in the field: effectiveness and behavior of a unique prey defense. Journal of Experimental Biology. 215 *(24): 4278–4287. doi:10.1242/jeb.076943.*

75. Dunning, D. C.; Roeder, K. D. (1965).Moth Sounds and the Insect-Catching Behavior of Bats. Science. 147 *(3654): 173–174. doi:10.1126/science.147.3654.173.*

76. Dunning, Dorothy C. (1968).Warning Sounds of moths. Zeitschrift für Tierpsychologie. 25: 125–138. doi:10.1111/j.1439-0310.1968.tb00008.

77. Roeder, Kenneth D. (1967). Nerve cells and insect behavior (Rev. ed., 1. Harvard Univ. Press paperback ed.). Cambridge, Mass [u.a.]: Harvard Univ.

Press. ISBN 0674608011.

78. Fullard, James H.; Fenton, M. Brock; Simmons, James A. (1979).Jamming bat echolocation: the clicks of arctiid moths. Canadian Journal of Zoology. 57 *(3): 647–649.* doi:10.1139/z79-076.

79. Fullard, JH; Simmons, JA; Saillant, PA (1994).Jamming bat echolocation: the dogbane tiger moth Cycnia tenera times its clicks to the terminal attack calls of the big brown bat Eptesicus fuscus.. *The Journal of Experimental Biology.* 194 *(1): 285–98. PMID 7964403.*

80. Surlykke, Annemarie; Miller, Lee A. (1985).The influence of arctiid moth clicks on bat echolocation; jamming or warning?. *Journal of Comparative Physiology A.* 156 *(6): 831–843.* doi:10.1007/BF00610835.

81. Ratcliffe, J. M.; J. H. Fullard (2005).The adaptive function of tiger moth clicks against echolocating bats: an experimental and synthetic approach. Journal of Experimental Biology. 208 *(24): 4689–4698. doi:10.1242/jeb.01927. PMID 16326950.*

82. Miller, Lee A. (1991).Arctiid moth clicks can degrade the accuracy of range difference discrimination in echolocating big brown bats, Eptesicus fuscus. Journal of Comparative Physiology A. 168 *(5): 571–579.* doi:10.1007/BF00215079.

83. Tougaard, J.; Casseday, J. H.; Covey, E. (1998).Arctiid moths and bat echolocation: broad-band clicks interfere with neural responses to auditory stimuli in the nuclei of the lateral lemniscus of the big brown bat. Journal of Comparative Physiology A. 182 *(2): 203–215.* doi:10.1007/s003590050171.

84. Szewczak, J. M.; E. Arnett (2007).Field Test Results of a Potential Acoustic Deterrent to Reduce Bat Mortality from Wind Turbines (PDF). Unpublished Report.

CHAPTER 7

Observational Learning

Chapter Contents

Observational learning describes the process of learning through watching others, retaining the information, and then later replicating the behaviors that were observed.

INTRODUCTION

Observational learning is learning that occurs through observing the behavior of others. It is a form of social learning which takes various forms, based on various processes. In humans, this form of learning seems to not need reinforcement to occur, but instead, requires a social model such as a parent, sibling, friend, or teacher. Particularly in childhood, a model is someone of authority or higher status. In animals, observational learning is often based on classical conditioning, in which an instinctive behavior is elicited by observing the behavior of another (e.g. mobbing in birds), but other processes may be involved as well.

HUMAN OBSERVATIONAL LEARNING

Many behaviors that a learner observes, remembers, and imitates are actions that models display, even though the model may not intentionally try to instill a particular behavior. A child may learn to swear, smack, smoke, and deem other inappropriate behavior acceptable through poor modeling. Bandura claims that children continually learn desirable and undesirable behavior through observational learning. Observational learning suggests that an individual's environment, cognition, and behavior all integrate and ultimately determine how the individual functions.

Through observational learning, individual behaviors can spread across a culture through a process called *diffusion chain*. This basically occurs when an individual first learns a behavior by observing another individual and that individual serves as a model through whom other individuals learn the behavior, and so on.

Culture plays a role in whether observational learning is the dominant learning style in a person or community. Some cultures expect children to actively participate in their communities and are therefore exposed to different trades and roles on a daily basis. This exposure allows children to observe and learn the different skills and practices that are valued in their communities.

Albert Bandura, who is known for the classic Bobo doll experiment, identified this basic form of learning in 1961. The importance of observational learning lies in helping individuals, especially children, acquire new responses by observing others' behavior.

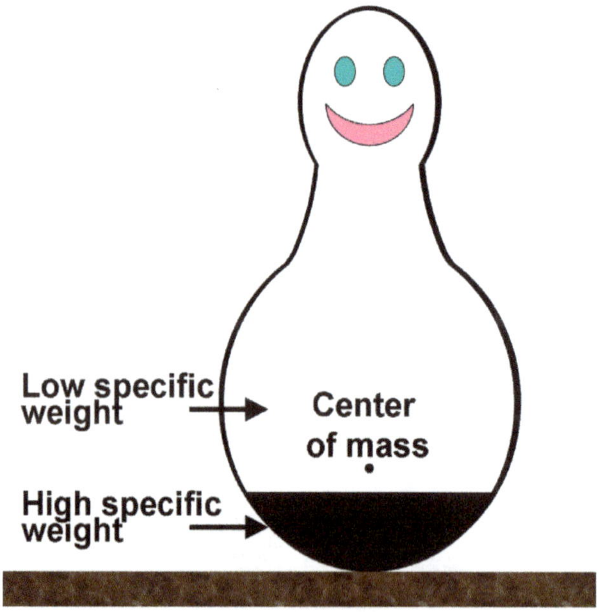

Bobo doll.

Albert Bandura states that people's behavior could be determined by their environment. Observational learning occurs through observing negative and positive behaviors. Bandura believes in reciprocal determinism in which the environment can influence people's behavior and vice versa. For instance, the Bobo doll experiment shows that model, in a determined environment, affects children's behavior. In this experiment Bandura demonstrates that one group of children placed in an aggressive environment would act the same way, while the control group and the other group of children placed in a passive role model environment hardly shows any type of aggression.

In communities where children's primary mode of learning is through observation, the children are rarely separated from adult activities. This incorporation into the adult world at an early age allows children to use observational learning skills in multiple spheres of life. This learning through observation requires keen attentive abilities. Culturally, they learn that their participation and contributions are valued in their communities. This teaches children that it is their duty, as members of the community, to observe others' contributions so they gradually become involved and participate further in the community.

STAGES

Bandura's social cognitive learning theory states that there are four stages involved in observational learning:

- **Attention**: Observers cannot learn unless they pay attention to what's happening around them. This process is influenced by characteristics of the model, such as how much one likes or identifies with the model, and by characteristics of the observer, such as the observer's expectations or level of emotional arousal.
- **Retention/Memory:** Observers must not only recognize the observed behavior but also remember it at some later time. This process depends on the observer's ability to code or structure the information in an easily remembered form or to mentally or physically rehearse the model's actions.
- **Initiation/Motor**: Observers must be physically and/intellectually capable of producing the act. In many cases the observer possesses the necessary responses. But sometimes, reproducing the model's actions may involve skills the observer has not yet acquired. It is one thing to carefully watch a circus juggler, but it is quite another to go home and repeat those acts.

- **Motivation**: Coaches also give pep talks, recognizing the importance of motivational processes to learning.

Bandura clearly distinguishes between learning and performance. Unless motivated, a person does not produce learned behavior. This motivation can come from external reinforcement, such as the experimenter's promise of reward in some of Bandura's studies, or the bribe of a parent. Or it can come from vicarious reinforcement, based on the observation that models are rewarded. High-status models can affect performance through motivation. For example, girls aged 11 to 14 performed better on a motor performance task when they thought it was demonstrated by a high-status cheerleader than by a low-status model.

Some have even added a step between attention and retention involving encoding a behavior.

Observational learning leads to a change in an individual's behavior along three dimensions:

- An individual thinks about a situation in a different way and may have incentive to react to it.
- The change is a result of a person's direct experiences as opposed to being in-born.
- For the most part, the change an individual has made is permanent.

EFFECT ON BEHAVIOR

Learning to play Djembe.

According to Bandura's social cognitive learning theory, observational learning can affect behavior in many ways, with both positive and negative consequences. It can teach completely new behaviors, for one. It can also increase or decrease the frequency of behaviors that have previously been learned. Observational learning can even encourage behaviors that were previously forbidden (for example, the violent behavior towards the Bobo doll that children imitated in Albert Bandura's study). Observational learning can also influence behaviors that are similar to, but not identical to, the ones being modeled. For example, seeing a model excel at playing the piano may motivate an observer to play the saxophone.

Age Difference

Albert Bandura stressed that developing children learn from different social models, meaning that no two children are exposed to exactly the same modeling influence. From infancy to adolescence, they are exposed to various social models. A 2013 study found that a toddlers' previous social familiarity with a model was not always necessary for learning and that they were also able to learn from observing a stranger demonstrating or modeling a new action to another stranger. It was once believed that babies could not imitate actions until the latter half of the first year. However a number of studies now report that infants as young as seven days can imitate simple facial expressions. By the latter half of their first year, 9-month-old babies can imitate actions hours after they first see them. As they continue to develop, toddlers around age two can acquire important personal and social skills by imitating a social model.

Deferred imitation is an important developmental milestone in a two-year-old, in which children not only construct symbolic representations, but can also remember information. Unlike toddlers, children of elementary school age are less likely to rely on imagination to represent an experience. Instead, they can verbally describe the model's behavior. Since this form of learning does not need reinforcement, it is more likely to occur regularly.

As age increases, age-related observational learning motor skills may decrease in athletes and golfers. Younger and skilled golfers have higher observational learning compared to older golfers and less skilled golfers.

Observational causal Learning

Humans use observational causal learning to watch other people's actions and use the information gained to find out how something works and how we can do it ourselves.

A study of 25-month-old infants found that they can learn causal relations from observing human interventions. They also learn by observing normal actions not created by intentional human action.

Comparisons with Imitation

Observational learning is presumed to have occurred when an organism copies an improbable action or action outcome that it has observed and the matching behavior cannot be explained by an alternative mechanism. Psychologists have been particularly interested in the form of observational learning known as imitation and in how to distinguish imitation from other processes. To successfully make this distinction, one must separate the degree to which behavioral similarity results from (a) predisposed behavior, (b) increased motivation resulting from the presence of another animal, (c) attention drawn to a place or object, (d) learning about the way the environment works, as distinguished from what we think of as (e) imitation (the copying of the demonstrated behavior) .

Observational learning differs from imitative learning in that it does not require a duplication of the behavior exhibited by the model. For example, the learner may observe an unwanted behavior and the subsequent consequences, and thus learn to refrain from that behavior. For example, Riopelle (1960) found that monkeys did better with observational learning if they saw the tutor monkey make a mistake before making the right choice. Heyes (1993) distinguished imitation and non-imitative social learning in the following way: imitation occurs when animals learn about behavior from observing conspecifics, whereas non-imitative social learning occurs when animals learn about the environment from observing others.

Not all imitation and learning through observing is the same, and they often differ in the degree to which they take on an active or passive form. John Dewey describes an important distinction between two different forms of imitation: imitation as an end in itself and imitation with a purpose. Imitation as an end is more akin to mimicry, in which a person copies another's act to repeat that action again. This kind of imitation is often observed in animals. Imitation with a purpose utilizes the imitative act as a means to accomplish something more significant. Whereas the more passive form of imitation as an end has been documented in some European American communities, the other kind of more active, purposeful imitation has been documented in other communities around the world.

Observation may take on a more active form in children's learning in multiple Indigenous American communities. Ethnographic anthropological studies in Yucatec Mayan and Quechua Peruvian communities provide evidence that the home or community-centered economic systems of these cultures allow children to witness first-hand, activities that are meaningful to their own livelihoods and the overall well-being of the community. These children have the opportunity to observe activities that are relevant within the context of that community, which gives them a reason to sharpen their attention to the practical knowledge they are exposed to. This does not mean that they have to observe the activities even though they are present. The children often make an active decision to stay in attendance while a community activity is taking place to observe and learn. This decision underscores the significance of this learning style in many indigenous American communities. It goes far beyond learning mundane tasks through rote imitation; it is central to children's gradual transformation into informed members of their communities' unique practices. There was also a study, done with children, that concluded that Imitated behavior can be recalled and used in another situation or the same.

Apprenticeship

Apprenticeship can involve both observational learning and modelling. Apprentices gain their skills in part through working with masters in their profession and through observing and evaluating the work of their fellow apprentices. Examples include renaissance inventor/painter Leonardo da Vinci and Michelangelo, before succeeding in their profession they were apprentices.

Learning without Imitation

Michael Tomasello described various ways of observational learning without the process of imitation in animals (ethology): Exposure- Individuals learn about theisr environment with a close proximity to other individuals that have more experience. For example, a young dolphin learning the location of a plethora of fish by staying near its mother.

- Stimulus enhancement - Individuals become interested in an object from watching others interact with it. Increased interest in an object may result in object manipulation, which facilitates new object-related behaviors by trial-and-error learning. For example, a young killer whale might become interested in playing with a sea lion pup after watching other whales toss the sea lion pup around. After playing with the pup, the killer whale may develop foraging behaviors appropriate to such prey. In this case, the killer whale did not learn to prey on sea lions by observing other whales do so, but rather the killer whale became intrigued after observing other whales play with the pup. After the killer whale became interested, then its interactions with the sea lion resulted in behaviors that provoked future foraging efforts.
- Goal emulation-Individuals are enticed by the end result of an observed behavior and attempt the same outcome but with a different method. For example, Haggerty (1909) devised an experiment in which a monkey climbed up the side of a cage, stuck its arm into a wooden chute, and pulled a rope in the chute to release food. Another monkey was provided an opportunity to obtain the food after watching a monkey go through this process on four separate occasions. The monkey performed a different method and finally succeeded after trial and error.

Peer model Influences

Observational learning is very beneficial when there are positive, reinforcing peer models involved. Although individuals go through four different stages for observational learning: attention; retention ; production; and motivation, this does not simply mean that when an individual's attention is captured that it automatically sets the process in that exact order. One of the most important ongoing stages for observational learning, especially among children, is motivation and positive reinforcement.

Performance is enhanced when children are positively instructed on how they can improve a situation and where children actively participate alongside a more skilled person. Examples of this are scaffolding and guided participation. Scaffolding refers to an expert responding contingently to a novice so the novice gradually increases their understanding of a problem. Guided participation refers to an expert actively engaging in a situation with a novice so the novice participates with or observes the adult to understand how to resolve a problem.

CULTURAL VARIATION

Cultural variation can be seen in the extent of information learned or absorbed by children through the use of observation and more specifically the use of observation without verbal requests for further information. For example, children from Mexican heritage families tend to learn and make better use of information observed during classroom demonstration then children of European heritage. Children of European heritage experience the type of learning that separates them from their family and community activities. They instead participate in lessons and other exercises in special settings such as school. Cultural backgrounds differ from each other in which children display certain characteristics in regards to learning an activity. Another example is seen in the immersion, of children in some Indigenous communities of the Americas, into the adult world and the effects it has on observational learning and the ability to complete multiple tasks simultaneously. This might be due to children in these communities having the opportunity to see a task being completed by their elders or peers and then trying to emulate the task. In doing so they learn to value observation and the skill-building it affords them because of the value it holds within their community. This type of observation is not passive, but reflects the child's intent to participate or learn within a community.

Observational learning can be seen taking place in many domains of Indigenous communities. The classroom setting is one significant example, and it functions differently for these communities in comparison to what is commonly present in Western schooling. The emphasis of keen observation in favor of supporting participation in ongoing activities strives to aid children to learn the important tools and ways of their community). Engaging in shared endeavors - with both the experienced and inexperienced - allows for the experienced to understand what the inexperienced need in order to grow in regards to the assessment of observational learning. The involvement of the inexperienced, or the children in this matter, can either be furthered by the children's learning or advancing into the activity performed by the assessment of observational learning. For the Indigenous communities to rely on observational learning is a way allowing for their children to be a part of ongoing activities in the community (Tharp, 2006).

Although learning in the IAC is not always the central focus when participating in an activity studies have shown that attention in intentional observation differs from accidental observation. Intentional participation is "keen observation and listening in anticipation of, or in the process of engaging in endeavors". This means that when they have the intention of participating in an event, their attention is more focused on the details, compared to when they are accidentally observing.

Observational learning can be an active process in many Indigenous American communities. The learner must take initiative to attend to activities going on around them. Children in these communities also take initiative to contribute their knowledge in ways that will benefit their community. For example, in many Indigenous American cultures, children perform household chores without being instructed to do so by adults. Instead, they observe a need for their contributions and take initiative to accomplish the tasks based on observations of others having done them. The learner's intrinsic motivations play an important role in the child's understanding and construction of meaning in these educational experiences. The independence and responsibility associated with observational learning in many Indigenous American communities are significant reasons why this method of learning can involve more than just watching and imitating. A learner must be actively engaged with their demonstrations and experiences in order to fully comprehend and apply the knowledge they obtain.

Indigenous Communities of the Americas

Mayan villagers

Children from indigenous heritage communities of the Americas often learn through observation, a strategy that can carry over into adulthood. The heightened value towards observation allows children to multi-task and actively engage in simultaneous activities. The exposure to an uncensored adult lifestyle allows children to observe and learn the skills and practices that are valued in their communities. Children observe elders, parents, and siblings complete tasks and learn to participate in them. They are seen as contributors and learn to observe multiple tasks being completed at once and can learn to complete a task, while still engaging with other community members without being distracted.

Indigenous communities provide more opportunities to incorporate children in everyday life. This can be seen in some Mayan communities where children are given full access to community events, which allows observational learning to occur more often. Other children in Mazahua, Mexico are known to intensely observe ongoing activities. In native northern Canadian and indigenous Mayan communities, children often learn as third-party observers from stories and conversations by others. Most young Mayan children are carried on their mother's back, allowing them to observe their mother's work and see the world as their mother sees it. Children are often allowed to learn without restrictions and with minimal guidance. They are self-motivated to learn and finish their chores. These children act as a second set of eyes and ears for their parents updating them about the community.

Children aged 6 to 8 in an indigenous heritage community in Guadalajara, Mexico participated in hard work, such as cooking or running errands, to benefit the whole family, while those in the city of Guadalajara rarely did so. These children participated more in adult regulated activities and had little time to play, while those from the indigenous-heritage community had more time to play and initiate in their own after-school activities.

Within certain indigenous communities people do not typically seek out explanation beyond basic observation. This is because they are competent in learning through astute observation. In a Guatemalan footloom factory amateur adult weavers observed skilled weavers over the course of weeks without questioning or being given explanations; the amateur weaver moved at their own pace and began when they felt confident. The framework of learning how to weave through observation can serve as a model that groups within a society use as a reference to guide their actions in particular domains of life. Communities that participate in observational learning promote tolerance and mutual understand of those coming from different cultural backgrounds.

Other human and animal behavior experiments

When an animal is given a task to complete, they are almost always more successful after observing another animal doing the same task before them. Experiments have been conducted on several different species with the same effect: animals can learn behaviors from peers. However, there is a need to distinguish the propagation of behavior and the stability of behavior. Research has shown that social learning can spread a behavior, but there are more factors regarding how a behavior carries across generations of an animal culture.

Learning in Fish

Experiments with ninespine sticklebacks showed that individuals will use social learning to locate food.

Social Learning in Pigeons

A study in 1996 at the University of Kentucky used a foraging device to test social learning in pigeons. A pigeon could access the food reward by either pecking at a treadle or stepping on it. Significant correspondence was found between the methods of how the observers accessed their food and the methods the initial model used in accessing the food.

Pigeon

Acquiring Foraging Niches

Studies have been conducted at the University of Oslo and University of Saskatchewan regarding the possibility of social learning in birds, delineating the difference between cultural and genetic acquisition. Strong evidence already exists for mate choice, bird song, predator recognition, and foraging.

Researchers cross-fostered eggs between nests of blue tits and great tits and observed the resulting behavior through audio-visual recording. Tits raised in the foster family learned their foster family's foraging sites early. This shift—from the sites the tits would among their own kind and the sites they learned from the foster parents—lasted for life. What young birds learn from foster parents, they eventually transmitted to their own offspring. This suggests cultural transmissions of foraging behavior over generations in the wild.

Social learning in crows

The University of Washington studied this phenomenon with crows, acknowledging the evolutionary tradeoff between acquiring costly information firsthand and learning that information socially with less cost to the individual but at the risk of inaccuracy. The experimenters exposed wild crows to a unique "dangerous face" mask as they trapped, banded, and released 7-15 birds at five different study places around Seattle, WA. An immediate scolding response to the mask after trapping by previously captured crows illustrates that the individual crow learned the danger of that mask. There was a scolding from crows that were captured that had not been captured initially. That response indicates conditioning from the mob of birds that assembled during the capture.

Horizontal social learning (learning from peers) is consistent with the lone crows that recognized the dangerous face without ever being captured. Children of captured crow parents were conditioned to scold the dangerous mask, which demonstrates vertical social learning (learning from parents). The crows that were captured directly had the most precise discrimination between dangerous and neutral masks than the crows that learned from the experience of their peers. The ability of crows to learn doubled the frequency of scolding, which spread at least 1.2 km from where the experiment started to over a 5-year period at one site.

Propagation of Animal Culture

Researchers at the Département d'Etudes Cognitives, Institut Jean Nicod, Ecole Normale Supérieure acknowledged a difficulty with research in social learning. To count acquired behavior as cultural, two conditions need must be met: the behavior must spread in a social group, and that behavior must be stable across generations. Research has provided evidence that imitation may play a role in the propagation of a behavior, but these researchers believe the fidelity of this evidence is not sufficient to prove stability of animal culture.

Other factors like ecological availability, reward-based factors, content-based factors, and source-based factors might explain the stability of animal culture in a wild rather than just imitation. As an example of ecological availability, chimps may learn how to fish for ants with a stick from their peers, but that behavior is also influenced by the particular type of ants as well as the condition. A behavior may be learned socially, but the fact that it was learned socially does not necessarily mean it will last. The fact that the behavior is rewarding has a role in cultural stability as well. The ability for socially-learned behaviors to stabilize across generations is also mitigated by the complexity of the behavior. Different individuals of a species, like crows, vary in their ability to use a complex tool. Finally, a behavior's stability in animal culture depends on the context in which they learn a behavior. If a behavior has already been adopted by a majority, then the behavior is more likely to carry across generations out of a need for conforming.

Animals are able to acquire behaviors from social learning, but whether or not that behavior carries across generations requires more investigation.

Hummingbird Experiment

Experiments with hummingbirds provided one example of apparent observational learning in a non-human organism. Hummingbirds were divided into two groups. Birds in one group were exposed to the feeding of a knowledgeable "tutor" bird; hummingbirds in the other group did not have this exposure. In subsequent tests the birds that had seen a tutor were more efficient feeders than the others.

Bottlenose Dolphin

Herman (2002) suggested that bottlenose dolphins produce goal-emulated behaviors rather than imitative ones. A dolphin that watches a model place a ball in a basket might place the ball in the basket when asked to mimic the behavior, but it may do so in a different manner seen.

RHESUS MONKEY

Kinnaman (1902) reported that one rhesus monkey learned to pull a plug from a box with its teeth to obtain food after watching another monkey succeed at this task.

Fredman (2012) also performed an experiment on observational behavior. In experiment 1, human-raised monkeys observed a familiar human model open a foraging box using a tool in one of two alternate ways: levering or poking. In experiment 2, mother-raised monkeys viewed similar techniques demonstrated by monkey models. A control group in each population saw no model. In both experiments, independent coders detected which technique experimental subjects had seen, thus confirming social learning. Further analyses examined copying at three levels of resolution.

The human-raised monkeys exhibited the greatest learning with the specific tool use technique they saw. Only monkeys who saw the levering model used the lever technique, by contrast with controls and those who witnessed poking. Mother-reared monkeys instead typically ignored the tool and exhibited fidelity at a lower level, tending only to re-create whichever result the model had achieved by either levering or poking.

Nevertheless, this level of social learning was associated with significantly greater levels of success in monkeys witnessing a model than in controls, an effect absent in the human-reared population. Results in both populations are consistent with a process of canalization of the repertoire in the direction of the approach witnessed, producing a narrower, socially shaped behavioral profile than among controls who saw no model.

Light box Experiment

Pinkham and Jaswal (2011) did an experiment to see if a child would learn how to turn on a light box by watching a parent. They found that children who saw a parent use their head to turn on the light box tended to do the task in that manner, but children who had not seen the parent chose a more efficient way, using their hands.

Swimming skill Performance

When adequate practice and appropriate feedback follow demonstrations, increased skill performance and learning occurs. Lewis (1974) did a study of children who had a fear of swimming and observed how modelling and going over swimming practices affected their overall performance. The experiment spanned nine days, and included many steps. The children were first assessed on their anxiety and swimming skills. Then they were placed into one of three conditional groups and exposed to these conditions over a few days.

At the end of each day, all children participated in a group lesson. The first group was a control group where the children watched a short cartoon video unrelated to swimming. The second group was a peer mastery group, which watched a short video of similar-aged children who had very good task performances and high confidence. Lastly, the third group was a peer coping group, whose subjects watched a video of similar-aged children who progressed from low task performances and low confidence statements to high task performances and high confidence statements.

The day following the exposures to each condition, the children were reassessed. Finally, the children were also assessed a few days later for a follow up assessment. Upon reassessment, it was shown that the two model groups who watched videos of children similar in age had successful rates on the skills assessed because they perceived the models as informational and motivational.

NEUROSCIENCE

Recent research in neuroscience has implicated mirror neurons as a neurophysiological basis for observational learning. These specialized visuomotor neurons fire action potentials when an individual performs a motor task and also fire when an individual passively observes another individual performing the same motor task. In observational motor learning, the process begins with a visual presentation of another individual performing a motor task, this acts as a model. The learner then needs to transform the observed visual information into internal motor commands that will allow them to perform the motor task, this is known as visuomotor transformation. Mirror neuron networks provide a mechanism for visuo-motor and motor-visual transformation and interaction. Similar networks of mirror neurons have also been implicated in social learning, motor cognition and social cognition.

Social Cognitive Theory

Social cognitive theory (SCT), used in psychology, education, and communication, holds that portions of an individual's knowledge acquisition can be directly related to observing others within the context of social interactions, experiences, and outside media influences. The theory states that when people observe a model performing a behavior and the consequences of that behavior, they remember the sequence of events and use this information to guide subsequent behaviors. Observing a model can also prompt the viewer to engage in behavior they already learned. In other words, people do not learn new behaviors solely by trying them and either succeeding or failing, but rather, the survival of humanity is dependent upon the replication of the actions of others. Depending on whether people are rewarded or punished for their behavior and the outcome of the behavior, the observer may choose to replicate behavior modeled. Media provides models for a vast array of people in many different environmental settings.

The conceptual roots for social cognitive theory come from Edwin B. Holt and Harold Chapman Brown's 1931 book theorizing that all animal action is based on fulfilling the psychological needs of feeling, emotion, and desire . The most notable component of this theory is that it predicted a person cannot learn to imitate until they are imitated.

In 1941, Neal E. Miller and John Dollard presented their book with a revision of Holt's social learning and imitation theory. They argued four factors contribute to learning: drives, cues, responses, and rewards. One driver is social motivation, which includes imitativeness, the process of matching an act to an appropriate cue of where and when to perform the act. A behavior is imitated depending on whether the model receives a positive or negative response consequences. Miller and Dollard argued that if one were motivated to learn a particular behavior, then that particular behavior would be learned through clear observations. By imitating these observed actions the individual observer would solidify that learned action and would be rewarded with positive reinforcement.

The proposition of social learning was expanded upon and theorized by Canadian psychologist Albert Bandura. Bandura, along with his students and colleagues conducted a series of studies, known as the Bobo doll experiment, in 1961 and 1963 to find out why and when children display aggressive behaviors. These studies demonstrated the value of modeling for acquiring novel behaviors. These studies helped Bandura publish his seminal article and book in 1977 that expanded on the idea of how behavior is acquired, and thus built from Miller and Dollard's research. In Bandura's 1977 article, he claimed that Social Learning Theory shows a direct correlation between a person's perceived self-efficacy and behavioral change. Self-efficacy comes from four sources: performance accomplishments, vicarious experience, verbal persuasion, and physiological states .

In 1986, Bandura published his second book, which expanded and renamed his original theory. He called the new theory *social cognitive theory*. Bandura changed the name to emphasize the major role cognition plays in encoding and performing behaviors. In this book, Bandura argued that human behavior is caused by personal, behavioral, and environmental influences.

In 2001, Bandura brought SCT to mass communication in his journal article that stated the theory could be used to analyze how "symbolic communication influences human thought, affect and action". The theory shows how new behavior diffuses through society by psychosocial factors governing acquisition and adoption of the behavior.

SCT has been applied to many areas of human functioning such as career choice and organizational behavior as well as in understanding classroom motivation, learning, and achievement.

Current Status

Social Cognitive Theory originated in psychology but based on an unofficial November 2013 Google Scholar search only 2 percent of articles published on SCT are in the pure psychology field. About 20 percent of articles are from Education and 16 percent from Business. The majority of publications using SCT, 56 percent, come from the field of Applied Health Psychology. The majority of current research in Health Psychology focuses on testing SCT in behavioral change campaigns as opposed to expanding on the theory. Campaign topics include: increasing fruit and vegetable intake, increasing physical activity, HIV education, and breastfeeding.

Born 1925, Bandura is still influencing the world with expansions of SCT. His recent work, published May 2011, focuses on how SCT impacts areas of both health and population effects in relation to climate change. He proposes that these problems could be solved through television serial dramas that show models similar to viewers performing the desired behavior. Specifically on Health, Bandura writes that currently there is little incentive for doctors to write prescriptions for healthy behavior, but he believes the cost of fixing health problems start to outweigh the benefits of being healthy. Bandura argues that we are on the cusp of moving from a disease model (focusing on people with problems) to a health model (focusing on people being healthy) and SCT is the theory that should be used to further a healthy society. Specifically on Population, Bandura states population growth is a global crisis because of its correlation with depletion and degradation of our planet's resources. Bandura argues that SCT should be used to get people to use birth control, reduce gender inequality through education, and to model environmental conservation to improve the state of the planet.

Social cognitive theory is a learning theory based on the idea that people learn by observing others. These learned behaviors can be central to one's personality. While social psychologists agree that the environment one grows up in contributes to behavior, the individual person (and therefore cognition) is just as important. People learn by observing others, with the environment, behavior, and cognition all as the chief factors in influencing development in a reciprocal triadic relationship. For example, each behavior witnessed can change a person's way of thinking (cognition). Similarly, the environment one is raised in may influence later behaviors, just as a father's mindset (also cognition) determines the environment in which his children are raised.

The core concepts of this theory can be explained by Bandura's schematization of triadic reciprocal causation in his book chapter, The schema shows how the reproduction of an observed behavior is influenced by the interaction of the following three determinants:

- Personal: Whether the individual has high or low self-efficacy toward the behavior (i.e. Get the learner to believe in his or her personal abilities to correctly complete a behavior).
- Behavioral: The response an individual receives after they perform a behavior (i.e. Provide chances for the learner to experience successful learning as a result of performing the behavior correctly).
- Environmental: Aspects of the environment or setting that influence the individual's ability to successfully complete a behavior (i.e. Make environmental conditions conducive for improved self-efficacy by providing appropriate support and materials).

It is important to note that learning can occur without a change in behavior. According to J.E. Ormrod's general principles of social learning, while a visible change in behavior is the most common proof of learning, it is not absolutely necessary. Social learning theorists say that because people can learn through observation alone, their learning may not necessarily be shown in their performance.

Theoretical Components

Modeling

Social cognitive theory revolves around the process of knowledge acquisition or learning directly correlated to the observation of models. The models can be those of an interpersonal imitation or media sources. Effective modeling teaches general rules and strategies for dealing with different situations.

To illustrate that people learn from watching others, Albert Bandura and his colleagues constructed a series of experiments using a Bobo doll. In the first experiment, children were exposed to either an aggressive or non-aggressive model of either the same sex or opposite sex as the child. There was also a control group. The aggressive models played with the Bobo doll in an aggressive manner, while the non-aggressive models played with other toys. They found that children who were exposed to the aggressive models performed more aggressive actions toward the Bobo doll afterward, and that boys were more likely to do so than girls.

Following that study, Albert Bandura tested whether the same was true for models presented through media by constructing an experiment he called *Bobo Doll Behavior: A Study of Aggression*. In this experiment Bandura exposed a group of children to a video featuring violent and aggressive actions. After the video he then placed the children in a room with a Bobo doll to see how they behaved with it. Through this experiment, Bandura discovered that children who had watched the violent video subjected the dolls to more aggressive and violent behavior, while children not exposed to the video did not. This experiment displays the social cognitive theory because it depicts how people reenact behaviors they see in the media. In this case, the children in this experiment reenacted the model of violence they directly learned from the video.

Observations should include:

- *Attention* Observers selectively give attention to specific social behavior depending on accessibility, relevance, complexity, functional value of the behavior or some observer's personal attributes such as cognitive capability, value preference, preconceptions.
- *Retention* Observe a behavior and subsequent consequences, then convert that observation to a symbol that can be accessed for future reenactments of the behavior. Note: When a *positive behavior* is shown a positive reinforcement should follow, this parallel is similar for *negative behavior*.
- *Production* refers to the symbolic representation of the original behavior being translated into action through reproduction of the observed behavior in seemingly appropriate contexts. During reproduction of the behavior, a person receives feedback from others and can adjust their representation for future references.
- *Motivational process* reenacts a behavior depending on responses and consequences the observer receives when reenacting that behavior.

Modeling does not limit to only live demonstrations but also verbal and written behaviour can act as indirect forms of modeling. Modeling not only allows students to learn behaviour that they should repeat but also to inhibit certain behaviours. For instance, if a teacher glares at one student who is talking out of turn, other students may suppress this behavior to avoid a similar reaction. Teachers model both material objectives and underlying curriculum of virtuous living. Teachers should also be dedicated to the building of high self-efficacy levels in their students by recognizing their accomplishments.

Outcome Expectancies

To learn a particular behaviour, people must understand what the potential outcome is if they repeat that behaviour. The observer does not expect the actual rewards or punishments incurred by the model, but anticipates similar outcomes when imitating the behavior (called *outcome expectancies*), which is why modeling impacts cognition and behavior. These expectancies are heavily influenced by the environment that the observer grows up in; for example, the expected consequences for a DUI in the United States of America are a fine, with possible jail time, whereas the same charge in another country might lead to the infliction of the death penalty.

For example, in the case of a student, the instructions the teacher provides help students see what outcome a particular behaviour leads to. It is the duty of the teacher to teach a student that when a behaviour is successfully learned, the outcomes are meaningful and valuable to the students.

Self-Efficacy

Social cognitive theory posits that learning most likely occurs if there is a close identification between the observer and the model and if the observer also has a good deal of self-efficacy. Self–efficacy is the extent to which an individual believes that they can master a particular skill. Self-efficacy beliefs function as an important set of proximal determinants of human motivation, affect, and action—which operate on action through motivational, cognitive, and affective intervening processes.

According to Bandura, self-efficacy is "the belief in one's capabilities to organize and execute the courses of action required to manage prospective situations".Bandura and other researchers have found an individual's self-efficacy plays a major role in how goals, tasks, and challenges are approached. Individuals with high self-efficacy are more likely to believe they can master challenging problems and they can recover quickly from setbacks and disappointments. Individuals with low self-efficacy tend to be less confident and don't believe they can perform well, which leads them to avoid challenging tasks. Therefore, self-efficacy plays a central role in behavior performance. Observers who have high level of self-efficacy are more likely to adopt observational learning behaviors.

Self-efficacy can be developed or increased by:

- **Mastery experience**, which is a process that helps an individual achieve simple tasks that lead to more complex objectives.
- **Social modeling** provides an identifiable model that shows the processes that accomplish a behavior.
- **Improving physical and emotional states** refers to ensuring a person is rested and relaxed prior to attempting a new behavior. The less relaxed, the less patient, the more likely they won't attain the goal behavior.
- **Verbal persuasion** is providing encouragement for a person to complete a task or achieve a certain behavior.

For example, students become more effortful, active, pay attention, highly motivated and better learners when they perceive that they have mastered a particular task. It is the duty of the teacher to allow student to perceive in their efficacy by providing feedback to understand their level of proficiency. Teachers should ensure that the students have the knowledge and strategies they need to complete the tasks.

Self-efficacy has also been used to predict behavior in various health related situations such as weight loss, quitting smoking, and recovery from heart attack. In relation to exercise science, self-efficacy has produced some of the most consistent results revealing an increase in participation in exercise.

Identification

Identification allows the observer to feel a one-to-one similarity with the model, and can thus lead to a higher chance of the observer following through with the modeled action. People are more likely to follow behaviors modeled by someone with whom they can identify with. The more commonalities or emotional attachments perceived between the observer and the model, the more likely the observer learns and reenact the modeled behavior.

Applications

Social cognitive theory is applied today in many different areas—e.g., mass media, public health, education, and marketing. Examples of the theory in application include: Use of celebrities to endorse and introduce products to specific demographics, in which social cognitive theory encompasses all four domains. Miller's 2005 study found that choosing the proper gender, age, and ethnicity for models ensured the success of an AIDS campaign to inner city teenagers. This occurred because participants could identify with a recognizable peer, have a greater sense of self-efficacy, and then imitate the actions to learn the proper preventions and actions. A study by Azza Ahmed in 2009 looked to see if there would be an increase in breastfeeding by mothers of preterm infants when exposed to a breastfeeding educational program guided by SCT. Sixty mothers were randomly assigned to either participate in the program or they were given routine care. The program consisted of SCT strategies that touched on all three SCT determinants: personal – showing models performing breastfeeding correctly to improve self-efficacy, behavioral –weekly check-ins for three months reinforced participants' skills, environmental – mothers were given an observational checklist to make sure they successfully completed the behavior. The author found that mothers exposed to the program showed significant improvement in their breastfeeding skills, were more likely to exclusively breastfeed, and had fewer problems then the mothers who were not exposed to the educational program.

Morality

Social cognitive theory emphasizes a large difference between an individual's ability to be morally competent and morally performing. Moral competence involves having the ability to perform a moral behavior, whereas moral performance indicates actually following one's idea of moral behavior in a specific situation. Moral competencies include:

- what an individual is capable of
- what an individual knows
- what an individual's skills are

- an individual's awareness of moral rules and regulations
- an individual's cognitive ability to construct behaviors

As far as an individual's development is concerned, moral competence is the growth of cognitive-sensory processes; simply put, being aware of what is considered right and wrong. By comparison, moral performance is influenced by the possible rewards and incentives to act a certain way. For example, a person's moral competence might tell them that stealing is wrong and frowned upon by society; however, if the reward for stealing is a substantial sum, their moral performance might indicate a different line of thought. Therein lies the core of social cognitive theory.

For the most part, social cognitive theory remains the same for various cultures. Since the concepts of moral behavior did not vary much between cultures (as crimes like murder, theft, and unwarranted violence are illegal in virtually every society), there is not much room for people to have different views on what is morally right or wrong. The main reason that social cognitive theory applies to all nations is because it does not say what is moral and immoral; it simply states that we can acknowledge these two concepts. Our actions in real-life scenarios are based on whether we believe the action is moral and whether the reward for violating our morals is significant enough, and nothing else.

CETACEAN INTELLIGENCE

Cetacean intelligence refers to the cognitive capabilities of the Cetacea order of mammals. This order includes whales, porpoises, and dolphins.

Brain size

Brain size was previously considered a major indicator of the intelligence of an animal. However, many other factors also affect intelligence. Recent discoveries concerning bird intelligence have called into question the usefulness of brain size as an indicator. Since most of the brain is used for maintaining bodily functions, greater ratios of brain to body mass may increase the amount of brain mass available for more complex cognitive tasks Allometric analysis indicates that mammalian brain size scales at approximately the ⅔ or ¾ exponent of the body mass. Comparison of a particular animal's brain size with the expected brain size based on such allometric analysis provides an encephalization quotient (EQ) that can be used as another indication of the animal's intelligence.

- Sperm whales (*Physeter macrocephalus*) have the largest known brain mass of any extant animal, averaging 7.8 kg in mature males.
- Bottlenose dolphins (*Tursiops truncatus*) have an absolute brain mass of 1500-1700 grams. This is slightly greater than that of humans (1300-1400 grams) and about four times that of chimpanzees (400 grams).
- The brain to body mass ratio (as distinct from encephalization quotient) in some members of the odontocete superfamily Delphinoidea (dolphins, porpoises, belugas, and narwhals) is second only to modern humans, and greater than all other mammals (there is debate whether that of the treeshrew might be second). In some dolphins, it is less than half that of humans: 0.9% versus 2.1%. This

comparison seems more favorable if the large amount of blubber (15-20% of mass) that dolphins require for insulation is omitted.

- The encephalization quotient varies widely between species. The La Plata dolphin has an EQ of approximately 1.67; the Ganges River dolphin of 1.55; the orca of 2.57; the bottlenose dolphin of 4.14; and the tucuxi dolphin of 4.56; In comparison to other animals, elephants have an EQ ranging from 1.13 to 2.36; chimpanzees of approximately 2.49; dogs of 1.17; cats of 1.00; and mice of 0.50.
- The majority of mammals are born with a brain close to 90% of the adult weight. Humans are born with 28% of the adult weight, chimpanzees with 54%,bottlenose dolphins with 42.5%, and elephants with 35%.

The discovery of spindle cells (neurons without extensive branching, known also as "von Economo neurons , or VENs) in the brains of the humpback whale, fin whale, sperm whale, killer whale, bottlenose dolphins, Risso's dolphins, and beluga whales is another unique discovery. Humans, the great apes, and elephants, species all well known for their high intelligence, are the only others known to have spindle cells. Spindle neurons appear to play a central role in the development of intelligent behavior. Such a discovery may suggest a convergent evolution of these species.

Brain Structure

Elephant brains also show a similar complexity to dolphin brains, and are also more convoluted than that of humans, and with a cortex "thicker than that of cetaceans". It is generally agreed that the growth of the neocortex, both absolutely and relative to the rest of the brain, during human evolution, has been responsible for the evolution of human intelligence, however defined. While a complex neocortex usually indicates high intelligence, there are exceptions to this. For example, the echidna has a highly developed brain, yet is not widely considered to be very intelligent.

In 2015, it was shown for the first time that a species of dolphin, the long-finned pilot whale, has more neocortical neurons than any mammal studied to date including humans. All sleeping mammals, including dolphins, experience a stage known as REM sleep. Unlike terrestrial mammals, dolphin brains contain a paralimbic lobe, which may possibly be used for sensory processing. The dolphin is a voluntary breather, even during sleep, with the result that veterinary anaesthesia of dolphins is impossible, as it would result in asphyxiation. Ridgway reports that EEGs show alternating hemispheric asymmetry in slow waves during sleep, with occasional sleep-like waves from both hemispheres. This result has been interpreted to mean that dolphins sleep only one hemisphere of their brain at a time, possibly to control their voluntary respiration system or to be vigilant for predators. This is also given as explanation for the large size of their brains.

Dolphin brain stem transmission time is faster than that normally found in humans, and is approximately equivalent to the speed found in rats. As echo-location is the dolphin's primary means of sensing its environment – analogous to eyes in primates – and since sound travels four and a half times faster in water than in air, scientists speculate that the faster brain stem transmission time, and perhaps the paralimbic lobe as well, assist quicker processing of sound. The dolphin's greater dependence on sound processing is evident in the structure of its brain: its neural area devoted to visual imaging is only about one-tenth that of the human brain, while the area devoted to acoustical imaging is about 10 times that of the human brain. Sensory experiments suggest a great degree of cross-modal integration in the processing of shapes between echolocative and visual areas of the brain. Unlike the case of the human brain, the cetacean optic chiasm is completely crossed, and there is behavioral evidence for hemispheric dominance for vision.

Behavior

Pod Characteristics

Dolphin group sizes vary quite dramatically. River dolphins usually congregate in fairly small groups from 6 to 12 in number or, in some species, singly or in pairs. The individuals in these small groups know and recognize one another. Other species such as the oceanic Pantropical Spotted Dolphin, Common Dolphin and Spinner Dolphin travel in large groups of hundreds of individuals. It is unknown whether every member of the group is acquainted with every other. However, there is no doubt that such large packs can act as a single cohesive unit – observations show that if an unexpected disturbance, such as a shark approach, occurs from the flank or from beneath the group, the group moves in near-unison to avoid the threat. This means that the dolphins must be aware not only of their near neighbors but also of other individuals nearby – in a similar manner to which humans perform "Audience waves . This is achieved by sight, and possibly also echolocation. One speculative hypothesis proposed by Jerison (1986) is that members of a pod of dolphins are able to share echolocation results with each other to create a better understanding of their surroundings.

Resident orcas living in British Columbia, Canada, and Washington, United States live in extremely stable family groups. The basis of this social structure is the matriline, consisting of a mother and her offspring, who travel with her for life. Male orcas never leave their mothers' pods, while female offspring may branch off to form their own matriline if they have many offspring of their own. Males have a particularly strong bond with their mother, and travel with them their entire lives, which can exceed 50 years. It is interesting behavior, as it may seem that there would be no benefit from this except perhaps in hunting techniques, although they could join other groups to hunt. There are two interesting examples of this familial bond in males. Two male sons, identified as A38 and A39, constantly accompany their mother A30, despite the fact that she does not need protection and they can all hunt by themselves, and rarely leave her side. Researchers have noted that if one son wanders away, one always remains with the mother. Another example are the brothers A32, A37 and A46, whose mother (A36) died. Instead of the family disbanding, the three brothers remain constantly together.

Relationships in the orca population can be discovered through their vocalizations. Matrilines who share a common ancestor from only a few generations back share mostly the same dialect, comprising a pod. Pods who share some calls indicate a common ancestor from many generations back, and make up a clan. Interestingly, the orcas use these dialects to avoid in-breeding. They mate outside the clan, which is determined by the different vocalizations. There is evidence that other species of dolphins may also have dialects.

In bottlenose dolphin studies by Wells in Sarasota, Florida, and Smolker in Shark Bay, Australia, females of a community are all linked either directly or through a mutual association in an overall social structure known as *fission-fusion*. Groups of the strongest association are known as "bands", and their composition can remain stable over years. There is some genetic evidence that band members may be related, but these bands are not necessarily limited to a single matrilineal line. There is no evidence that bands compete with each other. In the same research areas, as well as in Moray Firth, Scotland, males form strong associations of two to three individuals, with a coefficient of association between 70 and 100. These groups of males are known as "alliances", and members often display synchronous behaviors such as respiration, jumping, and breaching. Alliance composition is stable on the order of tens of years, and may provide a benefit for the acquisition of females for mating. The complex social strategies of marine mammals such as bottlenose dolphins, "provide interesting parallels" with the social strategies of elephants and chimpanzees.[(p519)]

Complex Play

Dolphins are known to engage in complex play behavior, which includes such things as producing stable underwater toroidal air-core vortex rings or "bubble rings . There are two main methods of bubble ring production: rapid puffing of a burst of air into the water and allowing it to rise to the surface, forming a ring; or swimming repeatedly in a circle and then stopping to inject air into the helical vortex currents thus formed. The dolphin will often then examine its creation visually and with sonar. They also appear to enjoy biting the vortex-rings they've created, so that they burst into many separate normal bubbles and then rise quickly to the surface. Certain whales are also known to produce bubble rings, or even bubble-nets for the purpose of foraging. Many dolphin species are also known for playing by riding in waves, whether natural waves near the shoreline in a method akin to human body-surfing , or within the waves induced by the bow of a moving boat in a behavior known as *bow riding*.

Cross-species Cooperation

There have been instances in captivity of various species of dolphin and porpoise helping and interacting across species, including helping beached whales. Also they have been known to live alongside Resident (fish eating) Orca Whales for limited amounts of time. Dolphins have also been known to aid human swimmers in need, and in at least one instance a distressed dolphin approached human divers seeking assistance.

Creative Behavior

Aside from having exhibited the ability to learn complex tricks, dolphins have also demonstrated the ability to produce creative responses. This was studied by Karen Pryor during the mid-1960s at Sea Life Park in Hawaii, and was published as *The Creative Porpoise: Training for Novel Behavior* in 1969. The two test subjects were two rough-toothed dolphins (*Steno bredanensis*), named Malia (a regular show performer at Sea Life Park) and Hou (a research subject at adjacent Oceanic Institute). The experiment tested when and whether the dolphins would identify that they were being rewarded (with fish) for originality in behavior and was very successful. However, since only two dolphins were involved in the experiment, the study is difficult to generalize.

Starting with the dolphin named Malia, the method of the experiment was to choose a particular behavior exhibited by her each day and reward each display of that behavior throughout the day's session. At the start of each new day Malia would present the prior day's behavior, but only when a new behavior was exhibited was a reward given. All behaviors exhibited were, at least for a time, known behaviors of dolphins. After approximately two weeks Malia apparently exhausted "normal" behaviors and began to repeat performances. This was not rewarded.

According to Pryor, the dolphin became almost despondent. However, at the sixteenth session without novel behavior, the researchers were presented with a flip they had never seen before. This was reinforced. As related by Pryor, after the new display: "instead of offering that again she offered a tail swipe we'd never seen; we reinforced that. She began offering us all kinds of behavior that we hadn't seen in such a mad flurry that finally we could hardly choose what to throw fish at".

The second test subject, Hou, took thirty-three sessions to reach the same stage. On each occasion the experiment was stopped when the variability of dolphin behavior became too complex to make further positive reinforcement meaningful.

The same experiment was repeated with humans, and it took the volunteers about the same length of time to figure out what was being asked of them. After an initial period of frustration or anger, the humans realised they were being rewarded for novel behavior. In dolphins this realisation produced excitement and more and more novel behaviors – in humans it mostly just produced relief.

Captive orcas have often displayed interesting responses when they get 'bored' with activities. For instance, when Dr. Paul Spong worked with the orca Skana, he researched her visual skills. However, after performing favorably in the 72 trials per day, Skana suddenly began consistently getting every answer wrong. Dr Spong concluded that a few fish were not enough motivation. He began playing music, which seemed to provide Skana with much more motivation.

At the Institute for Marine Mammal Studies in Mississippi, it has also been observed that the resident dolphins seem to show an awareness of the future. The dolphins are trained to keep their own tank clean by retrieving rubbish and bringing it to a keeper, to be rewarded with a fish. However, one dolphin, named Kelly, has apparently learned a way to get more fish, by hoarding the rubbish under a rock at the bottom of the pool and bringing it up one small piece at a time.

Use of Tools

As of 1984, scientists have observed wild bottlenose dolphins in Shark Bay, Western Australia using a basic tool. When searching for food on the sea floor, many of these dolphins were seen tearing off pieces of sponge and wrapping them around their rostra, presumably to prevent abrasions and facilitate digging.

Communication

Whale song is the sounds made by whales and which is used for different kinds of communication.

Dolphins emit two distinct kinds of acoustic signals, which are called *whistles* and *clicks*:

- Clicks – quick broadband burst pulses – are used for echolocation, although some lower-frequency broadband vocalizations may serve a non-echolocative purpose such as communication; for example, the pulsed calls of Orcas. Pulses in a click train are emitted at intervals of ≈35–50 milliseconds, and in general these inter-click intervals are slightly greater than the round-trip time of sound to the target.
- Whistles – narrow-band frequency modulated (FM) signals – are used for communicative purposes, such as contact calls, the pod-specific dialects of resident Orcas, or the signature whistle of bottlenose dolphins.

There is strong evidence that some specific whistles, called *signature whistles*, are used by dolphins to identify and/or call each other; dolphins have been observed emitting both other specimens' signature whistles, and their own. A unique signature whistle develops quite early in a dolphin's life, and it appears to be created in an imitation of the signature whistle of the dolphin's mother. Imitation of the signature whistle seems to occur only among the mother and its young, and among befriended adult males.

Xitco reported the ability of dolphins to eavesdrop passively on the active echolocative inspection of an object by another dolphin. Herman calls this effect the "acoustic flashlight" hypothesis, and may be related to findings by both Herman and Xitco on the comprehension of variations on the pointing gesture, including human pointing, dolphin postural pointing, and human gaze, in the sense of a redirection of another individual's attention, an ability which may require theory of mind

The environment where dolphins live makes experiments much more expensive and complicated than for many other species; additionally, the fact that cetaceans can emit and hear sounds (which are believed to be their main means of communication) in a range of frequencies much wider than that of humans means that sophisticated equipment, which was scarcely available in the past, is needed to record and analyse them. For example, clicks can contain significant energy in frequencies greater than 110 kHz (for comparison, it is unusual for a human to be able to hear sounds above 20 kHz), requiring that equipment have a sampling rates of at least 220 kHz; MHz-capable hardware is often used.

In addition to the acoustic communication channel, the visual modality is also significant. The contrasting pigmentation of the body may be used, for example with "flashes" of the hypopigmented ventral area of some species, as can the production of bubble streams during signature whistling. Also, much of the synchronous and cooperative behaviors, as described in the Behavior section of this entry, as well as cooperative foraging methods, likely are managed at least partly by visual means.

While there is little evidence for dolphin language, experiments have shown that they can learn human sign language. Akeakamai, a bottlenose dolphin, was able to understand both individual words and basic sentences like touch the frisbee with your tail and then jump over it (Herman, Richards, & Wolz 1984). Dolphins have also exhibited the ability to understand the significance of the ordering of each set of tasks in one sentence.

A study conducted by the University of Chicago showed that bottlenose dolphins can remember whistles of other dolphins they'd lived with after 20 years of separation. Each dolphin has a unique whistle that functions like a name, allowing the marine mammals to keep close social bonds. The new research shows that dolphins have the longest memory yet known in any species other than people.

Self-Awareness

Self-awareness is seen, by some, to be a sign of highly developed, abstract thinking. Self-awareness, though not well-defined scientifically, is believed to be the precursor to more advanced processes like meta-cognitive reasoning (thinking about thinking) that are typical of humans. Scientific research in this field has suggested that bottlenose dolphins, alongside elephants and great apes, possess self-awareness.

A recent study conducted by the University of Chicago shows that bottlenose dolphins can remember whistles of other dolphins they'd lived with after 20 years of separation. Each dolphin has a unique whistle that functions like a name, allowing the marine mammals to keep close social bonds. This also shows that Dolphins see themselves and other Dolphins as distinct individuals. The new research shows that dolphins have the longest memory yet known in any species other than people.

The most widely used test for self-awareness in animals is the mirror test, developed by Gordon Gallup in the 1970s, in which a temporary dye is placed on an animal's body, and the animal is then presented with a mirror.

Some scientists still disagree with these findings, arguing that the results of these tests are open to human interpretation and susceptible to the Clever Hans effect. This test is much less definitive than when used for primates, because primates can touch the mark or the mirror, while dolphins cannot, making their alleged self-recognition behavior less certain. Critics argue that behaviors that are said to identify self-awareness resemble existing social behaviors, and so researchers could be misinterpreting social responses to another dolphin. The researchers counter-argue that the behaviors shown to evidence self-awareness are very different from normal responses to another dolphin, including paying significantly more attention to another dolphin than towards their mirror image. Whereas apes can merely touch the mark on themselves with their fingers, dolphins show less definitive behavior of self-awareness, twisting and turning themselves to observe the mark.

In 1995, Marten and Psarakos used television to test dolphin self-awareness. They showed dolphins real-time footage of themselves, recorded footage, and another dolphin. They concluded that their evidence suggested self-awareness rather than social behavior. While this particular study has not been repeated since then, dolphins have since passed the mirror test.

Comparative Cognition

Research of the comparative cognition of the dolphin is one of the primary methods of the investigation of cetacean intelligence.

Examples of cognitive abilities investigated in the dolphin include concept formation, sensory skills, and the use of mental representation of dolphins. Such research has been ongoing since the late 1970s, and includes the specific topics of: acoustic mimicry, behavioral mimicry (inter- and intra-specific), comprehension of novel sequences in an artificial language (including non-finite state grammars as well as novel anomalous sequences), memory, monitoring of self-behaviors (including reporting on these, as well as avoiding or repeating them), reporting on the presence and absence of objects, object categorization, discrimination and matching (identity matching to sample, delayed matching to sample, arbitrary matching to sample, matching across echolocation and vision, reporting that no identity match exists, etc.), synchronous creative behaviors between two animals, comprehension of symbols for various body parts, comprehension of the pointing gesture and gaze (as made by dolphins or humans), problem solving, echolocative eavesdropping, and more. Some researchers include Louis Herman, Mark Xitco, John Gory, Stan Kuczaj, Adam Pack, and many others.

While these are largely laboratory studies, field studies relating to dolphin and cetacean cognition are also relevant to the issue of intelligence, including those proposing tool use, culture, fission-fusion social structure (including tracking alliances and other cooperative behavior), acoustic behavior (bottlenose dolphin signature whistles, sperm whale clicks, orca pod vocalizations), foraging methods (partial beaching, cooperation with human fishermen, herding fish into a ball, etc.). See: Richard Connor, Hal Whitehead, Peter Tyack, Janet Mann, Randall Wells, Kenneth Norris, B. Wursig, John Ford, Louis Herman, Diana Reiss, Lori Marino, Sam Ridgway, Paul Nachtigall, Eduardo Mercado, Denise Herzing, Whitlow Au.

In contrast to the primates, cetaceans are particularly far-removed from humans in evolutionary time. Therefore, cognitive abilities generally cannot be claimed to derive from a common ancestor, whereas such claims are sometimes made by researchers studying primate cognition. Though cetaceans and humans (in common with all mammals) had a common ancestor in the distant past, it was almost certainly of distinctly inferior cognitive abilities compared to its modern descendants. The early divergence of the human – dolphin ancestry line creates a problem in what cognitive tasks to test for because human/ dolphin brains have evolved so differently, with completely different cognitive abilities favoring their very different environments. Therefore, an anthropomorphic problem exists with exactly what cognitive abilities to test, how to test them, as well as the validity of the experimental results because of the completely different evolutionary lineage and environment human and cetaceans have. It was for this reason Dr. John C. Lilly proposed that developing a means of communicating with dolphins is necessary to have any future hope of communicating with an extraterrestrial organism of equal-or-greater intelligence to man, which also would have evolved in a different environment and evolutionary lineage.

NEUROSCIENCE AND INTELLIGENCE

From Wikipedia, the free encyclopedia

Neuroscience and intelligence refers to the various neurological factors that are partly responsible for the variation of intelligence within a species or between different species. A large amount of research in this area has been focused on the neural basis of human intelligence. Historic approaches to study the neuroscience of intelligence consisted of correlating external head parameters, for example head circumference, to intelligence. Post-mortem measures of brain weight and brain volume have also been used. More recent methodologies focus on examining correlates of intelligence within the living brain using techniques such as magnetic resonance imaging (MRI), functional MRI (fMRI), electroencephalography (EEG), positron emission tomography and other non-invasive measures of brain structure and activity.

Researchers have been able to identify correlates of intelligence within the Brain and its functioning. These include overall brain volume, grey matter volume,white matter volume, white matter integrity, cortical thickness and Neural Efficiency. Although the evidence base for our understanding of the neural basis of human intelligence has increased greatly over the past 30 years, even more research is needed to fully understand it.

The neural basis of intelligence has also been examined in animals such as primates, cetaceans and rodents.

Humans

Brain Volume

One of the main methods used to establish a relationship between intelligence and the brain is to use measures of Brain volume. The earliest attempts at estimating brain volume were done using measures of external head parameters, such as head circumference, however, such approximations proved to be inaccurate when estimating brain size. More recent methodologies that were employed to study this relationship are post-mortem measures of brain weight and volume. However, such measures also proved to be somewhat inconclusive, yielding diverging results depending on sex, which hemisphere was examined and on the type of intelligence measured.

The most widespread methodology in contemporary neuroscience to measure brain volume and size is MRI. MRI is a non-invasive technique used to study the brain structure and function (using fMRI) of living subjects. Overall, larger brain size and volume is associated with better cognitive functioning and higher intelligence. The correlations range from 0.0 to as high as 0.6, and are predominantly positive. The specific regions that show the most robust correlation between volume and intelligence are the frontal, temporal and parietal lobes of the brain. Therefore, it can be safely concluded that larger brains predict greater intelligence.

However, researchers have cautioned against oversimplifying this view. A meta-analytic review by McDaniel found that the correlation between Intelligence and in vivo brain size was larger for females (0.40) than for males (0.25). The same study also found that the correlation between brain size and Intelligence differed for age within sex, with children showing smaller correlations. Furthermore, the hypothesis has been put forward that the relationship between larger brain volumes and higher intelligence is facilitated not by the global increase of brain volume, but instead by the enlargement of selective parts of the brain associated with specific tasks. For example, monolingual adolescents learning new words is displayed growth in gray matter density in bilateral posterior supramarginal gyri directly related to the number of words learned. Similarly, learning to juggle increased grey matter volume in the occipito-temporal cortex for subjects who could not juggle previously, indicating that brain volume is dependent on a large variety of things and not a perfect measure for intelligence.

Another point of caution is that while larger brain volume is associated with higher intelligence, this relationship only explains a modest amount of variance within the distribution of intelligence itself. The correlations reported between brain volume, brain size and intelligence only explain 12% to 36% of the variance in the distribution of intelligence. The amount of variance explained by brain volume also depends on the type of intelligence measured. Up to 36% of variance in verbal intelligence can be explained by brain volume, while only approximately 10% of variance in visuospatial intelligence can be explained by brain volume.These caveats imply that there are other major factors influencing how intelligent an individual is apart from brain size. In a large meta-analysis consisting of 88 studies Pietschnig et al. (2015) estimated the correlation between brain volume and intelligence to be about correlation coefficient of 0.24 which equates to 6% variance. Researcher Jakob Pietschnig showed that this strength of the positive association of brain volume and IQ has been overestimated in the literature, but still remains robust. He has stated that *It is tempting to interpret this association in the context of human cognitive evolution and species differences in brain size and cognitive ability, we show that it is not warranted to interpret brain size as an isomorphic proxy of human intelligence differences*". Another 2015 study by researcher Stuart J Ritchie found that brain size explained 12% of the variance in intelligence among individuals.

Grey Matter

Grey matter has been examined as a potential biological foundation for differences in intelligence. Similarly to brain volume, global grey matter volume is positively associated with intelligence. More specifically, higher intelligence has been associated with larger cortical grey matter in the prefrontal and posterior temporal cortex in adults. Furthermore, both verbal and nonverbal intelligence have been shown to be positively correlated with grey matter volume across the parietal, temporal and occipital lobes in young healthy adults, implying that intelligence is associated with a wide variety of structures within the brain.

There appear to be sex differences between the relationship of grey matter to intelligence between men and women. Men appear to show more intelligence to grey matter correlations in the frontal and parietal lobes, while the strongest correlations between intelligence and grey matter in women can be found in the frontal lobes and Broca's Area. However, these differences do not seem to impact overall Intelligence, implying that the same cognitive ability levels can be attained in different ways.

One specific methodology used to study grey matter correlates of intelligence in areas of the brain is known as Voxel-based morphometry (VBM). VBM allows researchers to specify areas of interest with great spatial resolution, allowing the examination of grey matter areas correlated with intelligence with greater special resolution. VBM has been used to correlate grey matter positively with intelligence in the frontal, temporal, parietal, and occipital lobes in healthy adults. VBM has also been used to show that grey matter volume in the medial region of the prefrontal cortex and the dorsomedial prefrontal cortex correlate positively with intelligence in a group of 55 healthy adults. VBM has also been successfully used to establish a positive correlation between grey matter volumes in the anterior cingulate and intelligence in children aged 5 to 18 years old.

Grey matter has also been shown to positively correlate with intelligence in children. Reis and colleagues have found that grey matter in the prefrontal cortex contributes most robustly to variance in Intelligence in children between 5 and 17, while subcortical grey matter is related to intelligence to a lesser extent. Frangou and colleagues examined the relationship between grey matter and intelligence in children and young adults aged between 12 and 21, and found that grey matter in the orbitofrontal cortex, cingulate gyrus, cerebellum and thalamus was positively correlated to intelligence, while grey matter in the caudate nucleus is negatively correlated with intelligence. However, the relationship between grey matter volume and intelligence only develops over time, as no significant positive relationship can be found between grey matter volume and intelligence in children under 11.

An underlying caveat to research into the relationship of grey matter volume and intelligence is demonstrated by the hypothesis of neural efficiency. The findings that more intelligent individuals are more efficient at using their neurons might indicate that the correlation of grey matter to intelligence reflects selective elimination of unused synapses, and thus a better brain circuitry.

White Matter

Similar to grey matter, white matter has been shown to correlate positively with intelligence in humans. White matter consists mainly of myleneated neuronal axons, responsible for delivering signals between neurons. Significant correlations between intelligence and the corpus callosum have been found, as larger callosal areas have been positively correlated with cognitive performance. However, there appear to be differences in importance for white matter between verbal and nonverbal intelligence, as although both verbal and nonverbal measures of intelligence correlate positively with the size of the corpus callosum, the correlation for intelligence and corpus callosum size was larger (.47) for nonverbal measures than that for verbal measures (.18). Anatomical mesh-based geometrical modelling has also shown positive correlations between the thickness of the corpus callosum and Intelligence in healthy adults.

White matter integrity has also been found to be related to Intelligence. White matter tract integrity is important for information processing speed, and therefore reduced white matter integrity is related to lower intelligence. The effect of white matter integrity is mediate entirely through information processing speed. These findings indicate that the brain is structurally interconnected and that axonal fibres are integrally important for fast information process, and thus general intelligence.

Contradicting the findings described above, VBM failed to find a relationship between the corpus callosum and intelligence in healthy adults. This contradiction can be viewed to signify that the relationship between white matter volume and intelligence is not as robust as that of grey matter and intelligence.

Cortical Thickness

Cortical thickness has also been found to correlate positively with intelligence in humans. However, the rate of growth of cortical thickness is also related to intelligence. In early childhood, cortical thickness displays a negative correlation with intelligence, while by late childhood this correlation has shifted to a positive one. More intelligent children were found to develop cortical thickness more steadily and over longer periods of time than less bright children. Studies have found cortical thickness to explain 5% in the variance of intelligence among individuals.

Cortical Convolution

Cortical convolution has increased the folding of the brain's surface over the course of human evolution. It has been hypothesized that the high degree of cortical convolution may be a neurological substrate that supports some of the human brain's most distinctive cognitive abilities. Consequently, individual intelligence within the human species might be modulated by the degree of cortical convolution.

Neural Efficiency

The neural efficiency hypothesis postulates that more intelligent individuals display less activation in the brain during cognitive tasks, as measured by Glucose metabolism. A small sample of participants (N=8) displayed negative correlations between intelligence and absolute regional metabolic rates ranging from -0.48 to -0.84, as measured by PET scans, indicating that brighter individuals were more effective processors of information, as they use less energy. According to an extensive review by Neubauer & Fink a large number of studies (N=27) have confirmed this finding using methods such as PET scans, EEG and fMRI.However, evidence contradicting the neural efficiency hypothesis also exists.

fMRI and EEG studies have revealed that task difficulty is an important factor affecting neural efficiency. More intelligent individuals display neural efficiency only when faced with tasks of subjectively easy to moderate difficulty, while no neural efficiency can be found during difficult tasks. In fact, more able individuals appear to invest more cortical resources in tasks of high difficulty. This appears to be especially true for the Prefrontal Cortex, as individuals with higher intelligence displayed increased activation of this area during difficult tasks compared to individuals with lower intelligence. It has been proposed that the main reason for the neural efficiency phenomenon could be that individuals with high intelligence are better at blocking out interfering information than individuals with low intelligence.

Further Research

Some scientists prefer to look at more qualitative variables to relate to the size of measurable regions of known function, for example relating the size of the primary visual cortex to its corresponding functions, that of visual performance.

In a study of the head growth of 633 term-born children from the Avon Longitudinal Study of Parents and Children cohort, it was shown that prenatal growth and growth during infancy were associated with subsequent IQ. The study's conclusion was that the brain volume a child achieves by the age of 1 year helps determine later intelligence. Growth in brain volume after infancy may not compensate for poorer earlier growth.

There is an association between IQ and myopia. One suggested explanation is that one or several pleiotropic gene(s) affect the size of the neocortex part of the brain and eyes simultaneously.

Parieto-Frontal Integration Theory

In 2007, Behavioral and Brain Sciences published a target article that put forth a biological model of intelligence based on 37 peer-reviewed neuroimaging studies (Jung & Haier, 2007). Their review of a wealth of data from functional imaging (functional magnetic resonance imaging and positron emission tomography) and structural imaging (diffusion MRI, voxel-based morphometry, in vivo magnetic resonance spectroscopy) argues that human intelligence arises from a distributed and integrated neural network comprising brain regions in the frontal and parietal lobes.

A recent lesion mapping study conducted by Barbey and colleagues provides evidence to support the P-FIT theory of intelligence.

Brain injuries at an early age isolated to one side of the brain typically results in relatively spared intellectual function and with IQ in the normal range.

Primates

Brain size

Another theory of brain size in vertebrates is that it may relate to social rather than mechanical skill. Cortical size relates directly to a pairbonding life style and among primates cerebral cortex size varies directly with the demands of living in a large complex social network.

Health

Several environmental factors related to health can lead to significant cognitive impairment, particularly if they occur during pregnancy and childhood when the brain is growing and the blood–brain barrier is less effective. Developed nations have implemented several health policies regarding nutrients and toxins known to influence cognitive function. These include laws requiring fortification of certain food products and laws establishing safe levels of pollutants (e.g. lead, mercury, and organochlorides). Comprehensive policy recommendations targeting reduction of cognitive impairment in children have been proposed.

SWARM INTELLIGENCE

Swarm intelligence (**SI**) is the collective behavior of decentralized, self-organized systems, natural or artificial. The concept is employed in work on artificial intelligence. The expression was introduced by Gerardo Beni and Jing Wang in 1989, in the context of cellular robotic systems.

SI systems consist typically of a population of simple agents or boids interacting locally with one another and with their environment. The inspiration often comes from nature, especially biological systems. The agents follow very simple rules, and although there is no centralized control structure dictating how individual agents should behave, local, and to a certain degree random, interactions between such agents lead to the emergence of "intelligent" global behavior, unknown to the individual agents. Examples in natural systems of SI include ant colonies, bird flocking, animal herding, bacterial growth, fish schooling and microbial intelligence.

The application of swarm principles to robots is called swarm robotics, while ‹swarm intelligence' refers to the more general set of algorithms. ‹Swarm prediction' has been used in the context of forecasting problems.

Models of Swarm Behavior

Boids (Reynolds 1987)

Boids is an artificial life program, developed by Craig Reynolds in 1986, which simulates the flocking behaviour of birds. His paper on this topic was published in 1987 in the proceedings of the ACM SIGGRAPH conference. The name "boid corresponds to a shortened version of bird-oid object , which refers to a bird-like object.

As with most artificial life simulations, Boids is an example of emergent behavior; that is, the complexity of Boids arises from the interaction of individual agents (the boids, in this case) adhering to a set of simple rules. The rules applied in the simplest Boids world are as follows:

separation: steer to avoid crowding local flockmates

alignment: steer towards the average heading of local flockmates

cohesion: steer to move toward the average position (center of mass) of local flockmates

More complex rules can be added, such as obstacle avoidance and goal seeking.

Self-Propelled Particles (Vicsek *et al.* 1995)

Self-propelled particles (SPP), also referred to as the *Vicsek model*, was introduced in 1995 by Vicsek *et al.* as a special case of the boids model introduced in 1986 by Reynolds. A swarm is modelled in SPP by a collection of particles that move with a constant speed but respond to a random perturbation by adopting at each time increment the average direction of motion of the other particles in their local neighbourhood. SPP models predict that swarming animals share certain properties at the group level, regardless of the type of animals in the swarm. Swarming systems give rise to emergent behaviours which occur at many different scales, some of which are turning out to be both universal and robust. It has become a challenge in theoretical physics to find minimal statistical models that capture these behaviours.

Metaheuristics

Evolutionary algorithms (EA), particle swarm optimization (PSO), ant colony optimization (ACO) and their variants dominate the field of nature-inspired metaheuristics. This list includes algorithms published up to circa the year 2000. A large number of more recent metaphor-inspired metaheuristics have started to attract criticism in the research community for hiding their lack of novelty behind an elaborate metaphor. For algorithms published since that time, see List of metaphor-based metaheuristics.

Stochastic Diffusion Search (Bishop 1989)

Stochastic diffusion search (SDS) is an agent-based probabilistic global search and optimization technique best suited to problems where the objective function can be decomposed into multiple independent partial-functions. Each agent maintains a hypothesis which is iteratively tested by evaluating a randomly selected partial objective function parameterised by the agent's current hypothesis. In the standard version of SDS such partial function evaluations are binary, resulting in each agent becoming active or inactive. Information on hypotheses is diffused across the population via inter-agent communication. Unlike the stigmergic communication used in ACO, in SDS agents communicate hypotheses via a one-to-one communication strategy analogous to the tandem running procedure observed in Leptothorax acervorum. A positive feedback mechanism ensures that, over time, a population of agents stabilise around the global-best solution. SDS is both an efficient and robust global search and optimisation algorithm, which has been extensively mathematically described. Recent work has involved merging the global search properties of SDS with other swarm intelligence algorithms.

Ant Colony Optimization (Dorigo 1992)

Ant colony optimization (ACO), introduced by Dorigo in his doctoral dissertation, is a class of optimization algorithms modeled on the actions of an ant colony. ACO is a probabilistic technique useful in problems that deal with finding better paths through graphs. Artificial 'ants'—simulation agents—locate optimal solutions by moving through a parameter space representing all possible solutions. Natural ants lay down pheromones directing each other to resources while exploring their environment. The simulated 'ants' similarly record their positions and the quality of their solutions, so that in later simulation iterations more ants locate better solutions.

Particle Swarm Optimization (Kennedy, Eberhart & Shi 1995)

Particle swarm optimization (PSO) is a global optimization algorithm for dealing with problems in which a best solution can be represented as a point or surface in an n-dimensional space. Hypotheses are plotted in this space and seeded with an initial velocity, as well as a communication channel between the particles.Particles then move through the solution space, and are evaluated according to some fitness criterion after each timestep. Over time, particles are accelerated towards those particles within their communication grouping which have better fitness values. The main advantage of such an approach over other global minimization strategies such as simulated annealing is that the large number of members that make up the particle swarm make the technique impressively resilient to the problem of local minima.

Applications

Swarm Intelligence-based techniques can be used in a number of applications. The U.S. military is investigating swarm techniques for controlling unmanned vehicles. The European Space Agency is thinking about an orbital swarm for self-assembly and interferometry. NASA is investigating the use of swarm technology for planetary mapping. A 1992 paper by M. Anthony Lewis and George A. Bekey discusses the possibility of using swarm intelligence to control nanobots within the body for the purpose of killing cancer tumors. Conversely al-Rifaie and Aber have used stochastic diffusion search to help locate tumours. Swarm intelligence has also been applied for data mining.

Ant-Based Routing

The use of swarm intelligence in telecommunication networks has also been researched, in the form of ant-based routing. This was pioneered separately by Dorigo et al. and Hewlett Packard in the mid-1990s, with a number of variations since. Basically, this uses a probabilistic routing table rewarding/reinforcing the route successfully traversed by each "ant" (a small control packet) which flood the network. Reinforcement of the route in the forwards, reverse direction and both simultaneously have been researched: backwards reinforcement requires a symmetric network and couples the two directions together; forwards reinforcement rewards a route before the outcome is known (but then one would pay for the cinema before one knows how good the film is). As the system behaves stochastically and is therefore lacking repeatability, there are large hurdles to commercial deployment. Mobile media and new technologies have the potential to change the threshold for collective action due to swarm intelligence (Rheingold: 2002, P175).

The location of transmission infrastructure for wireless communication networks is an important engineering problem involving competing objectives. A minimal selection of locations (or sites) are required subject to providing adequate area coverage for users. A very different-ant inspired swarm intelligence algorithm, stochastic diffusion search

(SDS), has been successfully used to provide a general model for this problem, related to circle packing and set covering. It has been shown that the SDS can be applied to identify suitable solutions even for large problem instances.

Airlines have also used ant-based routing in assigning aircraft arrivals to airport gates. At Southwest Airlines a software program uses swarm theory, or swarm intelligence—the idea that a colony of ants works better than one alone. Each pilot acts like an ant searching for the best airport gate. "The pilot learns from his experience what's the best for him, and it turns out that that's the best solution for the airline, Douglas A. Lawson explains. As a result, the "colony" of pilots always go to gates they can arrive at and depart from quickly. The program can even alert a pilot of plane back-ups before they happen. "We can anticipate that it's going to happen, so we'll have a gate available," Lawson says.

Crowd Simulation

Artists are using swarm technology as a means of creating complex interactive systems or simulating crowds.

Stanley and Stella in: Breaking the Ice was the first movie to make use of swarm technology for rendering, realistically depicting the movements of groups of fish and birds using the Boids system. Tim Burton's *Batman Returns* also made use of swarm technology for showing the movements of a group of bats. *The Lord of the Rings* film trilogy made use of similar technology, known as Massive, during battle scenes. Swarm technology is particularly attractive because it is cheap, robust, and simple.

Airlines have used swarm theory to simulate passengers boarding a plane. Southwest Airlines researcher Douglas A. Lawson used an ant-based computer simulation employing only six interaction rules to evaluate boarding times using various boarding methods. (Miller, 2010, xii-xviii).

Human Swarming

Enabled by mediating software such as the UNU collective intelligence platform, networks of distributed users can be organized into "human swarms" (also referred to as "social swarms") through the implementation of real-time closed-loop control systems. As published by Rosenberg (2015), such real-time control systems enable groups of human participants to behave as a unified collective intelligence. When logged into the UNU platform, for example, groups of distributed users can collectively answer questions, generate ideas, and make predictions as a singular emergent entity. Early testing shows that human swarms can out-predict individuals across a variety of real-world projections.

Swarm Grammars

Swarm grammars are swarms of stochastic grammars that can be evolved to describe complex properties such as found in art and architecture. These grammars interact as agents behaving according to rules of swarm intelligence. Such behavior can also suggest deep learning algorithms, in particular when mapping of such swarms to neural circuits is considered.

Swarmic Art

In a series of works al-Rifaie et al. have successfully used two swarm intelligence algorithms—one mimicking the behaviour of one species of ants (*Leptothorax acervorum*) foraging (stochastic diffusion search, SDS) and the other algorithm mimicking the behaviour of birds flocking (particle swarm optimization, PSO)—to describe a novel integration strategy exploiting the local search properties of the PSO with global SDS behaviour. The resulting hybrid algorithm is used to sketch novel drawings of an input image, exploiting an artistic tension between the local behaviour of the ‹birds flocking'—as they seek to follow the input sketch—and the global behaviour of the ants foraging —as they seek to encourage the flock to explore novel regions of the canvas. The creativity of this hybrid swarm system has been analysed under the philosophical light of the rhizome in the context of Deleuze's Orchid and Wasp metaphor.

In a more recent work of al-Rifaie et al., Swarmic Sketches and Attention Mechanism , introduces a novel approach deploying the mechanism of ‹attention' by adapting SDS to selectively attend to detailed areas of a digital canvas. Once the attention of the swarm is drawn to a certain line within the canvas, the capability of PSO is used to produce a ‹swarmic sketch' of the attended line. The swarms move throughout the digital canvas in an attempt to satisfy their dynamic roles—attention to areas with more details—associated to them via their fitness function. Having associated the rendering process with the concepts of attention, the performance of the participating swarms creates a unique, non-identical sketch each time the ‹artist' swarms embark on interpreting the input line drawings. In other works while PSO is responsible for the sketching process, SDS controls the attention of the swarm.

In a similar work, Swarmic Paintings and Colour Attention , non-photorealistic images are produced using SDS algorithm which, in the context of this work, is responsible for colour attention.

The computational creativity of the above-mentioned systems are discussed in through the two prerequisites of creativity (i.e. freedom and constraints) within the swarm intelligence's two infamous phases of exploration and exploitation.

Michael Theodore and Nikolaus Correll use swarm intelligent art installation to explore what it takes to have engineered systems to appear lifelike Notable work include Swarm Wall (2012) and endo-exo (2014).

REFERENCES

1. Shettleworth, S. J. "Cognition, Evolution, and Behavior", 2010 (2nd ed.) New York:Oxford,
2. Bandura,A. (1971) Psychological Modelling. New York: Lieber-Antherton
3. Schacter, Gilbert, & Wegner, D. L., D.T., & D. M. (2011). Psychology. Worth Publishers. p. 295.
4. Garton, A. F. (2007). Learning through collaboration: Is there a multicultural perspective?. AIP. pp. 195–216.
5. Hughes, Claire (2011). Hughes, Claire. (2011) Social Understanding and Social

Lives. New York, Ny: Psychology Press.

6. Most Human Behavior is learned Through Modeling .
7. Fleer, M. (2003). Early Childhood Education as an Evolving ‹Community of Practice' or as Lived ‹Social Reproduction': researching the ‹taken-for-granted' . Contemporary Issues in Early Childhood. **4** *(1): 64–79. doi:10.2304/ciec.2003.4.1.7.*
8. Bandura, Albert. Observational Learning. Learning and Memory. Ed. John H. Byrne. 2nd ed. New York: Macmillan Reference USA, 2004. 482-484. Gale Virtual Reference Library. Web. 6 Oct. 2014. Document URL http://go.galegroup.com/ps/i.do?id=GALE%7CCX3407100173&v=2.1&u=cuny_hunter&it=r&p=GVRL&sw=w&asid=06f2484b425a0c9f9606dff1b2a86c18
9. Weiss, Maureen R.; Ebbeck, Vicki; Rose, Debra J. (1992). Show and tell in the gymnasium revisited: Developmental differences in modeling and verbal rehearsal effects on motor skill learning and performance . Research quarterly for exercise and sport. **63** *(3): 292–301. doi:10.1080/02701367.1992.10608745.*
10. Weiss, Maureen et al. (1998). Observational Learning and the Fearful Child: Influence of Peer Models on Swimming Skill Performance and Psychological Responses. 380-394
11. Shimpi, Priya M.; Akhtar, Nameera; Moore, Chris (2013). Toddlers' Imitative Learning in Interactive and Observational Contexts: The Role of Age and Familiarity of the Model . Journal of Experimental Child Psychology. **116** *(2): 309–23. doi:10.1016/j.jecp.2013.06.008.*
12. Meltzoff, A (1988). Infants imitation after 1-week delay: Long -Term memory for novel acts and multiple stimuli . Developmental Psychology. **24**: 470–476. doi:10.1037/0012-1649.24.4.470.
13. Bandura,A. (1989). Social Cognitive Theory. In R. Vasta (ED.), Annals of child Development: Vol. 6. Theories of child development: Revised formulation and current issue (pp.1-60). Greenwich, CT : JAI Press
14. Law, Barbi; Hall, Craig (2009). The Relationships Among Skill Level, Age, and Golfers' Observational Learning Use . Sport Psychologist. **23** *(1): 42.*
15. Meltzoff, A. N.; Waismeyer, A.; Gopnik, A. (2012). Learning about causes from people: Observational causal learning in 24-month-old infants . Developmental Psychology. **48** *(5): 1215–1228.* doi:10.1037/a0027440.
16. Zentall, Thomas R (2012). Perspectives On Observational Learning In Animals . Journal of Comparative Psychology. **126** *(2): 114–128.* doi:10.1037/a0025381.
17. Riopelle, A.J. (1960). Observational learning of a position habit by monkeys . Journal of Comparative and Physiological Psychology. **53** *(5): 426–428.* doi:10.1037/h0046480.
18. Heyes, C. M. (1993). Imitation, culture and cognition . Animal Behaviour. **46**: 999–1010. doi:10.1006/anbe.1993.1281.
19. Dewey, John (1916). Democracy and Education. New York: Macmillan Co.
20. Gaskins, Paradise. The Anthropology of Learning in Childhood. Alta Mira

Press. pp. Chapter 5.

21. McLaughlin, L. J.; Brinley, J. F. (1973). Age and observational learning of a multiple-classification task . Developmental Psychology. **9** *(1): 9–15.* doi:10.1037/h0035069.

22. Groenendijk, Talita; Janssen, Tanja; Rijlaarsdam, Gert; Huub Van, Den Bergh (2013). Learning to Be Creative. The Effects of Observational Learning on Students' Design Products and Processes . Learning and Instruction. **28**: 35–47. doi:10.1016/j.learninstruc.2013.05.001.

23. Tomasello, M. (1999). The cultural origins of human cognition. Cambridge, MA: Harvard University Press. 248 pp.

24. Spence, K. W. (1937). Experimental studies of learning and higher mental processes in infra-human primates . Psychological Bulletin. **34**: 806–850. doi:10.1037/h0061498.

25. Haggerty, M. E. (1909). Imitation in monkeys . Journal of Comparative Neurology and Psychology. **19**: 337–455. doi:10.1002/cne.920190402. horizontal tab character in |journal= at position 38 (help)

26. Schaffer, David et al. (2010). Developmental Psychology, Childhood and Adolescence. 284

27. Cole, M. Culture and early childhood learning (PDF). Retrieved 15 November 2012.

28. Mejia-Arauz, R.; Rogoff, B.; Paradise, R. (2005). Cultural variation in children's observation during a demonstration . International Journal of Behavioral Development. **29**: 282–291. doi:10.1177/01650250544000062.

29. Rogoff, Barbara. Cultural Variation in Children's Attention and Learning. N.p.: n.p., n.d. N. pag. PsycINFO. Web.

30. Coppens, A., Alcala, L., Mejia-Arauz, R., Rogoff, B. (2014). Children's Initiative in Family Household Work in Mexico . Human Development. **57**: 116–130. doi:10.1159/000356768.

31. Gaskins, Suzanne. Open attention as a cultural tool for observational learning. (PDF). Kellogg Institute for International Studies University of Notre Dame. Retrieved 7 May 2014.

32. Rogoff, Barbara; Paradise, R.; Arauz, R.; Correa-Chavez, M. (2003). Firsthand learning through intent participation . Annual Review of Psychology. **54**: 175–203. doi:10.1146/annurev.psych.54.101601.145118.

33. Rogoff, Barbara; Paradise, Ruth; Correa-Chavez, M; Arauz, R (2003). Firsthand Learning through Intent Participation . Annual Review of Psychology. **54**: 175–203. doi:10.1146/annurev.psych.54.101601.145118.

34. Modiano,, Nancy (1973). Indian education in the Chiapas Highlands. New York: Holt, Rinehart and Winston. pp. 33–40. ISBN 0030842379.

35. Gaskins, Suzanne (Nov 1, 2000). Children's Daily Activities in a Mayan Village: A Culturally Grounded Description . Cross-Cultural Research. **34**: 375–389. doi:10.1177/106939710003400040.

36. Rogoff, Barbara; Mosier, Christine; Misty, Jayanthi; Göncü, Artin (Jan 1, 1989). Toddlers' Guided Participation in Cultural Activity . Cultural Dynamics. **2**: 209–237. doi:10.1177/092137408900200205.

37. Children's Initiative in Contributions to Family Work in Indigenous-Heritage and Cosmopolitan Communities in Mexico. (2014). 57(2-3).

38. Gee, J.; Green, J (1998). Discourse analysis, learning and social practice: A methodological study . Review of Research and Education.

39. Frith, Chris D., and Uta Frith. Mechanisms Of Social Cognition. Annual Review Of Psychology 2012; 63.: 287-313

40. Zentall, T. R.; Sutton, J. E.; Sherburne, L. M. (1996). True imitative learning in pigeons . Psychological Science. **7** *(6): 343–346. doi:10.1111/j.1467-9280.1996.tb00386.x.*

41. Slagsvold, Tore (2011). Social learning in birds and its role in shaping a foraging niche. (PDF). Retrieved Jan 18, 2016.

42. Slagsvold, T.; Wiebe, K. L. (2011). Social learning in birds and its role in shaping a foraging niche . Philosophical Transactions of the Royal Society B: Biological Sciences. **366** *(1567): 969–977. doi:10.1098/rstb.2010.0343.*

43. Cornell, H. N., Marzluff, J. M., & Pecoraro, S. (2012). Social learning spreads knowledge about dangerous humans among american crows. Proceedings of The Royal Society B: Biological Sciences,

44. Claidiere, N.; Sperber, D. (2010). Imitation explains the propagation, not the stability of animal culture . Proceedings of The Royal Society B: Biological Sciences. **277** *(1681): 651–659. doi:10.1098/rspb.2009.1615.*

45. Altshuler, D.; Nunn, A. (2001). Obeservational learning in hummingbirds . The Auk. **118** *(3): 795–799.* doi:10.2307/4089948.

46. Herman, L. M. (2002). Vocal, social, and self-imitation by bottlenosed dolphins. In K. Dautenhahn & C. Nehaniv (Eds.), Imitation in animals and artifacts (pp. 63–108). Cambridge: MIT Press.

47. Kinnaman, A. J. (1902). Mental life of two Macacus rhesus monkeys in captivity. The American Journal of Psychology, 13, 173–218.

48. Fredman, Tamar, and Andrew Whiten. Observational Learning from Tool using Models by Human-Reared and Mother-Reared Capuchin Monkeys (Cebus Apella). Animal Cognition 11.2 (2008): 295–309. ProQuest Research Library. Web. 30 Oct. 2012.

49. Pinkham, A.M.; Jaswal, V.K. (2011). Watch and learn? Infants privilege efficiency over pedagogy during imitative learning . Infancy. **16** *(5): 535–544. doi:10.1111/j.1532-7078.2010.00059.x.*

50. Weiss, Maureen et al. (1998). Observational Learning and the Fearful Child: Influence of Peer Models n Swimming Skill Performance and Psychological Responses. 380–394

51. Lago-Rodríguez, A.; Cheeran, B.; Koch, G.; Hortobagy, T.; Fernandez-del-Olmo, M. (2014). The role of mirror neurons in observational motor learning:

an integrative review . European Journal of Human Movement. **32**: 82–103.

52. Rizzolatti, G.; Fogassi, L. (2014). The mirror mechanism: recent findings and perspectives . Philosophical Transactions of the Royal Society B: Biological Sciences. **369** *(1644): 20130420. doi:10.1098/rstb.2013.0420.*

53. Jeannerod, M.; Arbib, M. A.; Rizzolatti, G.; Sakata, H. (1995). Grasping objects: the cortical mechanisms of visuomotor transformation . Trends in Neurosciences. **18** *(7): 314–320.* doi:10.1016/0166-2236(95)93921-j.

54. Uddin, L. Q.; Iacoboni, M.; Lange, C.; Keenan, J. P. (2007). The self and social cognition: the role of cortical midline structures and mirror neurons . Trends in Cognitive Sciences. **11** *(4): 153–157. doi:10.1016/j.tics.2007.01.001. PMID 17300981.*

55. Bandura, A., Social foundations of thought and action : a social cognitive theory. 1986, Englewood Cliffs, N.J.: Prentice-Hall.

56. Bandura, A. (2002). Social cognitive theory of mass communication. In J. Bryant & M. B. Oliver (Eds.), Media Effects: Advances in Theory and Research (pp. 94-124). New York, NY: Routledge.

57. Holt, E.B. & H.C. Brown (1931). Animal drive and the learning process, an essay toward radical empiricism. New York: H. Holt and Co.

58. Miller, N.E.; J. Dollard & R. Yale University (1941). Institute of Human, Social learning and imitation. New Haven; London: Pub. for the Institute of human relations by Yale university press; H. Milford, Oxford University Press.

59. Evans, R.I. & A. Bandura (1989). Albert Bandura, the man and his ideas—a dialogue. New York: Praeger.

60. Bandura, A (1977). Self-efficacy: Toward a Unifying Theory of Behavioral Change (PDF). Psychological Review. **84** *(2): 191–215. doi:10.1037/0033-295x.84.2.191. PMID 847061.*

61. Bandura, A. (2001). Social Cognitive Theory of Mass Communication (PDF). Media Psychology. **3** *(3): 265–299.* doi:10.1207/S1532785XMEP0303_03.

62. Lent, Robert; Steven D. Brown; Gail Hackett (August 1994). Toward a Unifying Social Cognitive Theory of Career and Academic Interest, Choice, and Performance . Journal of Vocational Behavior. **45** *(1): 79–122. doi:10.1006/jvbe.1994.1027.*

63. Bandura, A. (2011). The Social and Policy Impact of Social Cognitive Theory. In M. Mark, S. Donaldson, & B. Campbell (Eds.), Social Psychology and Evaluation. (pp. 33-70). New York, NY:Guilford Press.

64. Self-efficacy theory . Edutech Wiki.

65. Bandura, A. (1988). Organizational Application of Social Cognitive Theory (PDF). Australian Journal of Management. **13** *(2): 275–302.* doi:10.1177/031289628801300210.

66. Bandura, Albert; Ross, D.; Ross, S. (1961). Transmission of aggression through imitation of aggressive models . Journal of Abnormal and Social Psychology. **63**: 575–582. doi:10.1037/h0045925. PMID 13864605.

67. Bandura, Albert; Ross, D.; Ross, S. (1963). Imitation of film-mediated aggressive models . Journal of Abnormal and Social Psychology. **66**: 3–11. doi:10.1037/h0048687.

68. Bandura, A. (1989). Human Agency in Social Cognitive Theory (PDF). American Psychologist. **44** *(9): 1175–1184. doi:10.1037/0003-066X.44.9.1175. PMID 2782727.*

69. Bandura, A. Self-efficacy in changing societies. Cambridge; New York: Cambridge University Press.

70. McAlister AL, Perry CL, Parcel GS (2008). How Individuals, Environments, and Health Behaviors Interact: Social Cognitive Theory . Health Behavior and Health Education: Theory, Research, and Practice (4th ed.). San Francisco, CA: John Wiley & Sons, Inc. pp. 169–188.

71. Bandura, Albert (1993). Perceived Self Efficacy in Cognitive Development and Functioning . Educational Psychologist. **28** *(2): 117–148.* doi:10.1207/s15326985ep2802_3.

72. Weinberg, Robert S.; Gould, Daniel (2007). Foundation of Sport and Exercise Psychology (4th ed.). Champaign, IL: Human Kinetics. p. 422.

73. Miller, Katherine (2005). Communication Theories: Perspectives, Processes, and Contexts (2nd ed.). New York, New York: McGraw-Hill.

74. Ahmed, A. (2009). Effect of Breastfeeding Educational Program Based of [sic] Bandura Social Cognitive Theory on Breastfeeding Outcomes among Mothers of Preterm Infants. Midwest Nursing Research Society Conference. Accessed November 2011. http://hdl.handle.net/10755/160761

75. Santrock, J.W. (2008). A Topical Approach to Lifespan Development (M. Ryan, Ed., 4th ed.). New York, NY: McGraw-Hill Companies, Inc. (Original work published 2002), pp. 26, 30, 478

CHAPTER 8

Pain in Animals

Chapter Contents

In humans, pain is a distressing feeling often caused by intense or damaging stimuli. Whether animals apart from humans also experience pain is contentious. The standard measure of pain in humans is how a person reports that pain, (for example, on a pain scale). Only the person experiencing the pain can know the pain's quality and intensity, and the degree of suffering. However, for non-human animals, it is harder, if even possible, to know whether an emotional experience has occurred. Therefore, this concept is often excluded in definitions of **pain in animals**, such as that provided by Zimmerman: "an aversive sensory experience caused by actual or potential injury that elicits protective motor and vegetative reactions, results in learned avoidance and may modify species-specific behaviour, including social behaviour." Non-human animals cannot report their feelings to language-using humans in the same manner as human communication, but observation of their behaviour provides a reasonable indication as to the extent of their pain. Just as with doctors and medics who sometimes share no common language with their patients, the indicators of pain can still be understood.

A Galapagos shark hooked by a fishing boat

According to the U.S. National Research Council Committee on Recognition and Alleviation of Pain in Laboratory Animals, pain is experienced by many animal species, including mammals and possibly all vertebrates.

THE EXPERIENCE OF PAIN

Although there are numerous definitions of pain, almost all involve two key components. First, nociception is required. This is the ability to detect noxious stimuli which evoke a reflex response that rapidly moves the entire animal, or the affected part of its body, away from the source of the stimulus. The concept of nociception does not imply any adverse, subjective "feeling" – it is a reflex action. An example in humans would be the rapid withdrawal of a finger that has touched something hot – the withdrawal occurs before any sensation of pain is actually experienced.

The second component is the experience ofpainitself, or suffering – the internal, emotional interpretation of the nociceptive experience. Again in humans, this is when the withdrawn finger begins to hurt, moments after the withdrawal. Pain is therefore a private, emotional experience. Pain cannot be directly measured in other animals, including other humans; responses to putatively painful stimuli can be measured, but not the experience itself. To address this problem when assessing the capacity of other species to experience pain, argument-by-analogy is used. This is based on the principle that if an animal responds to a stimulus in a similar way to ourselves, it is likely to have had an analogous experience.

Nociception

Nociception usually involves the transmission of a signal along a chain of nerve fibers from the site of a noxious stimulus at the periphery to the spinal cord and brain. This process evokes a reflex arc response generated at the spinal cord and not involving the brain, such as flinching or withdrawal of a limb. Nociception is found, in one form or another, across all major animal taxa. Nociception can be observed using modern imaging techniques; and a physiological and behavioral response to nociception can be detected.

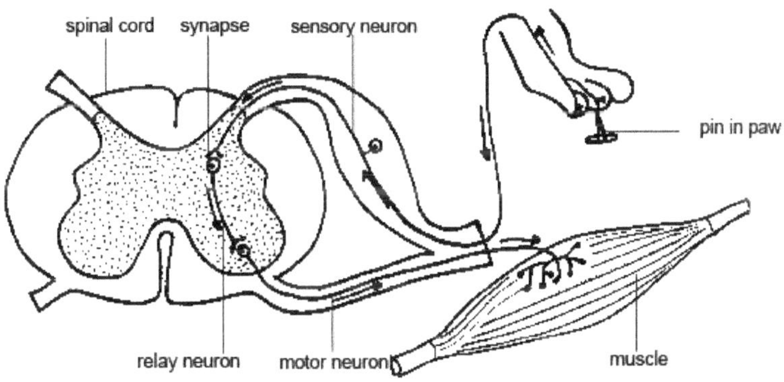

Reflex arc of a dog with a pin in her paw. Note there is no communication to the brain, but the paw is withdrawn by nervous impulses generated by the spinal cord. There is no conscious interpretation of the stimulus by the dog.

Pain

The nerve impulses of the nociception response may be conducted to the brain thereby registering the location, intensity, quality and unpleasantness of the stimulus. This subjective component of pain involves conscious awareness of both the sensation and the unpleasantness (the aversive, negative affect). The brain processes underlying conscious awareness of the unpleasantness (suffering), are not well understood.

Adaptive Value

The adaptive value of nociception is obvious; an organism detecting a noxious stimulus immediately withdraws the limb, appendage or entire body from the noxious stimulus and thereby avoids further (potential) injury. However, a characteristic of pain (in mammals at least) is that pain can result in hyperalgesia (a heightened sensitivity to noxious stimuli) and allodynia (a heightened sensitivity to non-noxious stimuli). When this heightened sensitisation occurs, the adaptive value is less clear. First, the pain arising from the heightened sensitisation can be disproportionate to the actual tissue damage caused. Second, the heightened sensitisation may also become chronic, persisting well beyond the tissues healing. This can mean that rather than the actual tissue damage causing pain, it is the pain due to the heightened sensitisation that becomes the concern. This means the sensitisation process is sometimes termed maladaptive. It is often suggested hyperalgesia and allodynia assist organisms to protect themselves during healing, but experimental evidence to support this has been lacking.

In 2014, the adaptive value of sensitisation due to injury was tested using the predatory interactions between longfin inshore squid (*Doryteuthis pealeii*) and black sea bass (*Centropristis striata*) which are natural predators of this squid. If injured squid are targeted by a bass, they began their defensive behaviours sooner (indicated by greater alert distances and longer flight initiation distances) than uninjured squid. If anaesthetic (1% ethanol and $MgCl_2$) is administered prior to the injury, this prevents the sensitisation and blocks the behavioural effect. The authors claim this study is the first experimental evidence to support the argument that nociceptive sensitisation is actually an adaptive response to injuries.

ARGUMENT-BY-ANALOGY

To address the problem of assessing the capacity of other species to experience the affective state of pain (to suffer), we resort to argument-by-analogy. This is based on the principle that if an animal responds to a stimulus in a similar way to ourselves, it is likely to have had an analogous experience. If we stick a pin in a chimpanzee's finger and she rapidly withdraws her hand, we use argument-by-analogy and infer that like us, she felt pain. If we are consistent, we should also infer a cockroach experiences the same when it writhes after being stuck with a pin. Analogous to humans, when given a choice of feeds, rats and chickens with clinical symptoms of pain will consume more of an analgesic-containing feed than animals not in pain. Additionally, the consumption of the analgesic carprofen in lame broiler chickens was positively correlated to the severity of lameness, and consumption resulted in an improved gait. Limitations of argument-by-analogy are that physical reactions may neither determine nor be motivated by mental states, and the approach is subject to criticism of anthropomorphic interpretation. For example, a single-celled organism such as an amoeba may writhe after being exposed to noxious stimuli despite the absence of nociception.

HISTORY

The idea that animals might not experience pain or suffering as humans do traces back at least to the 17th-century French philosopher, René Descartes, who argued that animals lack consciousness. Researchers remained unsure into the 1980s as to whether animals experience pain, and veterinarians trained in the U.S. before 1989 were simply taught to ignore animal pain. In his interactions with scientists and other veterinarians, Bernard Rollin was regularly asked to "prove" that animals are conscious, and to provide "scientifically acceptable" grounds for claiming that they feel pain. Some authors say that the view that animals feel pain differently is now a minority view. Academic reviews of the topic are more equivocal, noting that, although it is likely that some animals have at least simple conscious thoughts and feelings, some authors continue to question how reliably animal mental states can be determined.

IN DIFFERENT SPECIES

The ability to experience pain in an animal, or another human for that matter, cannot be determined directly but it may be inferred through analogous physiological and behavioral reactions. Although many animals share similar mechanisms of pain detection to those of humans, have similar areas of the brain involved in processing pain, and show similar pain behaviours, it is notoriously difficult to assess how animals actually experience pain.

Nociception

Nociceptive nerves, which preferentially detect (potential) injury-causing stimuli, have been identified in a variety of animals, including invertebrates. The medicinal leech, *Hirudo medicinalis,* and sea slug are classic model systems for studying nociception. Many other vertebrate and invertebrate animals also show nociceptive reflex responses similar to our own.

Pain

Many animals also exhibit more complex behavioural and physiological changes indicative of the ability to experience pain: they eat less food, their normal behaviour is disrupted, their social behaviour is suppressed, they may adopt unusual behaviour patterns, they may emit characteristic distress calls, experience respiratory and cardiovascular changes, as well as inflammation and release of stress hormones.

Some criteria that may indicate the potential of another species to feel pain include:

- Has a suitable nervous system and sensory receptors
- Physiological changes to noxious stimuli
- Displays protective motor reactions that might include reduced use of an affected area such as limping, rubbing, holding or autotomy
- Has opioid receptors and shows reduced responses to noxious stimuli when given analgesics and local anaesthetics
- Shows trade-offs between stimulus avoidance and other motivational requirements
- Shows avoidance learning
- High cognitive ability and sentience

VERTEBRATES

Fish

A typical human cutaneous nerve contains 83% C type trauma receptors (the type responsible for transmitting signals described by humans as excruciating pain); the same nerves in humans with congenital insensitivity to pain have only 24-28% C type receptors. The rainbow trout has about 5% C type fibres, while sharks and rays have 0%. Nevertheless, fish have been shown to have sensory neurons that are sensitive to damaging stimuli and are physiologically identical to human nociceptors. Behavioural and physiological responses to a painful event appear comparable to those seen in amphibians, birds, and mammals, and administration of an analgesic drug reduces these responses in fish.

Animal welfare advocates have raised concerns about the possible suffering of fish caused by angling. Some countries, e.g. Germany, have banned specific types of fishing, and the British RSPCA now formally prosecutes individuals who are cruel to fish.

INVERTEBRATES

Though it has been argued that most invertebrates do not feel pain, there is some evidence that invertebrates, especially the decapod crustaceans (e.g. crabs and lobsters) and cephalopods (e.g. octopuses), exhibit behavioural and physiological reactions indicating they may have the capacity for this experience. Nociceptors have been found in nematodes, annelids and molluscs. Most insects do not possess nociceptors, one known exception being the fruit fly. In vertebrates, endogenous opioids are neurochemicals that moderate pain by interacting with opiate receptors. Opioid peptides and opiate receptors occur naturally in nematodes, molluscs, insects and crustaceans. The presence of opioids in crustaceans has been interpreted as an indication that lobsters may be able to experience pain, although it has been claimedat present no certain conclusion can be drawn.

One suggested reason for rejecting a pain experience in invertebrates is that invertebrate brains are too small. However, brain size does not necessarily equate to complexity of function. Moreover, weight for body-weight, the cephalopod brain is in the same size bracket as the vertebrate brain, smaller than that of birds and mammals, but as big as or bigger than most fish brains.

Since September 2010, all cephalopods being used for scientific purposes in the EU are protected by EU Directive 2010/63/EU which states...there is scientific evidence of their [cephalopods] ability to experience pain, suffering, distress and lasting harm. In the UK, animal protection legislation means that cephalopods used for scientific purposes must be killed humanely, according to prescribed methods (known asSchedule 1 methods of euthanasia) known to minimise suffering.

IN MEDICINE AND RESEARCH

Veterinary medicine

Veterinary medicine uses, for actual or potential animal pain, the same analgesics and anesthetics as used in humans.

Dolorimetry

Dolorimetry (*dolor*: Latin: pain, grief) is the measurement of the pain response in animals, including humans. It is practiced occasionally in medicine, as a diagnostic tool, and is regularly used in research into the basic science of pain, and in testing the efficacy of analgesics. Non-human animal pain measurement techniques include the paw pressure test, tail flick test, hot plate test and grimace scales.

Laboratory Animals

Animals are kept in laboratories for a wide range of reasons, some of which may involve pain, suffering or distress, whilst others (e.g. many of those involved in breeding) will not. The extent to which animal testing causes pain and suffering in laboratory animals is the subject of much debate. Marian Stamp Dawkins defines "suffering" in laboratory animals as the experience of one of "a wide range of extremely unpleasant subjective (mental) states." The U.S. National Research Council has published guidelines on the care and use of laboratory animals, as well as a report on recognizing and alleviating pain in vertebrates.The United States Department of Agriculture defines a "painful procedure" in an animal study as one that would "reasonably be expected to cause more than slight or momentary pain or distress in a human being to which that procedure was applied." Some critics argue that, paradoxically, researchers raised in the era of increased awareness of animal welfare may be inclined to deny that animals are in pain simply because they do not want to see themselves as people who inflict it. PETA however argues that there is no doubt about animals in laboratories being inflicted with pain. In the UK, animal research likely to cause "pain, suffering, distress or lasting harm" is regulated by the Animals (Scientific Procedures) Act 1986 and research with the potential to cause pain is regulated by the Animal Welfare Act of 1966 in the US.

In the U.S., researchers are not required to provide laboratory animals with pain relief if the administration of such drugs would interfere with their experiment. Laboratory animal veterinarian Larry Carbone writes, "Without question, present public policy allows humans to cause laboratory animals unalleviated pain. The AWA, the *Guide for the Care and Use of Laboratory Animals*, and current Public Health Service policy all allow for the conduct of what are often called "Category E" studies – experiments in which animals are expected to undergo significant pain or distress that will be left untreated because treatments for pain would be expected to interfere with the experiment."

Severity Scales

Eleven countries have national classification systems of pain and suffering experienced by animals used in research: Australia, Canada, Finland, Germany, The Republic of Ireland, The Netherlands, New Zealand, Poland, Sweden, Switzerland, and the UK. The US also has a mandated national scientific animal-use classification system, but it is markedly different from other countries in that it reports on whether pain-relieving drugs were required and/or used. The first severity scales were implemented in 1986 by Finland and the UK. The number of severity categories ranges between 3 (Sweden and Finland) and 9 (Australia). In the UK, research projects are classified as "mild", "moderate", and "substantial" in terms of the suffering the researchers conducting the study say they may cause; a fourth category of "unclassified" means the animal was anesthetized and killed without recovering consciousness. It should be remembered that in the UK system, many research projects (e.g. transgenic breeding, feeding distasteful food) will require a license under the Animals (Scientific Procedures) Act 1986, but may cause little or no pain or suffering. In December 2001, 39 percent (1, 296) of project licenses in force were classified as "mild", 55 percent (1, 811) as "moderate", two percent (63) as "substantial", and 4 percent (139) as "unclassified". In 2009, of the project licenses issued, 35 percent (187) were classified as "mild", 61 percent (330) as "moderate", 2 percent (13) as "severe" and 2 percent (11) as unclassified.

In the US, the *Guide for the Care and Use of Laboratory Animals* defines the parameters for animal testing regulations. It states, "The ability to experience and respond to pain is widespread in the animal kingdom...Pain is a stressor and, if not relieved, can lead to unacceptable levels of stress and distress in animals.The *Guide* states that the ability to recognize the symptoms of pain in different species is essential for the people caring for and using animals. Accordingly, all issues of animal pain and distress, and their potential treatment with analgesia and anesthesia, are required regulatory issues for animal protocol approval.

CRUELTY TO ANIMALS

Cruelty to animals, also called **animal abuse** or **animal neglect**, is the intentional infliction by humans of suffering or harm upon any non-human animal, regardless of whether the act is against the law. More narrowly, it can be the causing of harm or suffering for specific gain, such as killing animals for food or for their fur; opinions differ about the extent of cruelty associated with a given method of slaughter. Cruelty to animals sometimes encompasses inflicting harm or suffering for personal amusement, as in zoosadism.

Laws concerning animal cruelty are designed to prevent needless cruelty. Divergent approaches to such laws occur in different jurisdictions throughout the world. For example, some laws govern methods of killing animals for food, clothing, or other products, and other laws concern the keeping of animals for entertainment, education, research, or pets. Cruelty to animals is not the same thing as disrespect towards animals.

In broad terms, there are three conceptual approaches to the issue of cruelty to animals. The animal welfare position holds that there is nothing inherently wrong with using animals for human purposes, such as food, clothing, entertainment, and research, but that it should be done in a way that minimizes unnecessary pain and suffering, sometimes referred to as "humane" treatment.

Utilitarian advocates argue from the position of costs and benefits and vary in their conclusions as to the allowable treatment of animals. Some utilitarians argue for a weaker approach which is closer to the animal welfare position, whereas others argue for a position that is similar to animal rights. Animal rights theorists criticize these positions, arguing that the words "unnecessary" and "humane" are subject to widely differing interpretations, and that animals have basic rights. They say that the only way to ensure protection for animals is to end their status as property and to ensure that they are never used as commodities.

Before modern biology revealed similarities between humans and animals, some thinkers considered humans completely distinct from animals; the controversy over animal welfare was virtually non-existent Certain thinkers, however, still viewed cruelty against animals as an injustice. Renaissance polymath Leonardo da Vinci's regard for animal welfare, for example, is well-documented. He was particularly troubled by the sight of birds in captivity, and (according to biographer Giorgio Vasari) once purchased caged birds in order to set them free. Da Vinci also expressed anger within his notebooks with the fact that humans use their strength and power to raise animals for slaughter.

Man beating a chained pitbull terrier with a strap. The strap is visible in the foreground.

René Descartes contrarily believed that non-humans are automata, complex machines with no soul, mind, or reason. In Cartesian dualism, consciousness was unique to human among all other animals and linked to physical matter by divine grace. However, close analysis shows that many human features such as complex sign usage, tool use, and self-consciousness can be found in some animals.

Charles Darwin, by presenting the theory of evolution, revolutionized the way that humans viewed their relationship with other species. Darwin believed that not only did human beings have a direct kinship with other animals, but the latter had social, mental and moral lives too. Later, in *The Descent of Man* (1871), he wrote: "There is no fundamental difference between man and the higher mammals in their mental faculties."

Some philosophers and intellectuals, such as Peter Singer and Tom Regan, have argued that animals' ability to feel pain as humans does make their well-being worthy of equal consideration. There are many precursors of this train of thought. Jeremy Bentham, the founder of utilitarianism, famously wrote in his *An Introduction to the Principles of Morals and Legislation* (1789):

> The question is not, can they reason nor can they talk? but, can they suffer?

These arguments have prompted some to suggest that animals' well-being should enter a social welfare function directly, not just indirectly via its effect only on human well-being.

In one survey of United States homeowners, 68% of respondents said they actually consider the price of meat a more important issue. Worldwide meat overconsumption is another factor that contributes to the miserable situation of farm animals.

Forms

Neglect

Animal cruelty can be broken down into two main categories: active and passive. Passive cruelty is typified by cases of neglect, in which the cruelty is a lack of action rather than the action itself. Examples of neglect are starvation, dehydration, parasite infestations, allowing a collar to grow into an animal's skin, inadequate shelter in extreme weather conditions, and failure to seek veterinary care when necessary. In many cases of neglect in which an investigator believes that the cruelty occurred out of ignorance, the investigator may attempt to educate the pet owner, then revisit the situation. In more severe cases, exigent circumstances may require that the animal be removed for veterinary care.

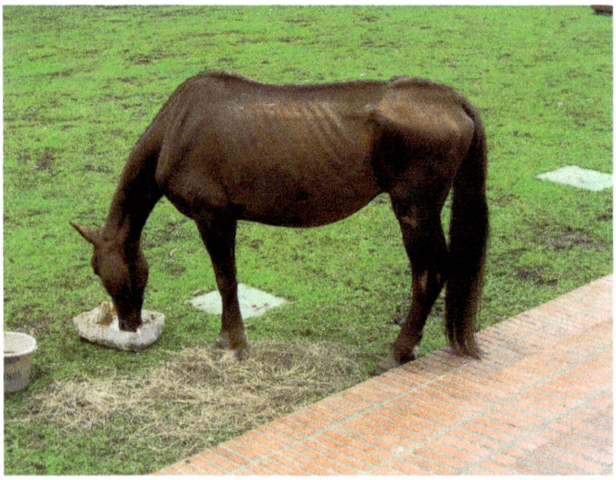

A starved and bruised horse eating at a veterinary clinic after rescue.

Industrial Animal Farming

Farm animals are generally produced in large, industrial facilities that house thousands of animals at high densities; these are sometimes called factory farms. The industrial nature of these facilities means that many routine procedures or animal husbandry practices impinge on the welfare of the animals and could arguably be considered as "cruelty". It has been suggested the number of animals hunted, kept as companions, used in laboratories, reared for the fur industry, raced, and used in zoos and circuses, is insignificant compared to farm animals, and therefore the "animal welfare issue" is numerically reducible to the "farm animal welfare issue". Similarly, it has been suggested by campaign groups that chickens, cows, pigs, and other farm animals are among the most numerous animals subjected to cruelty. For example, because male chickens do not lay eggs, newly hatched males are culled using macerators or grinders.

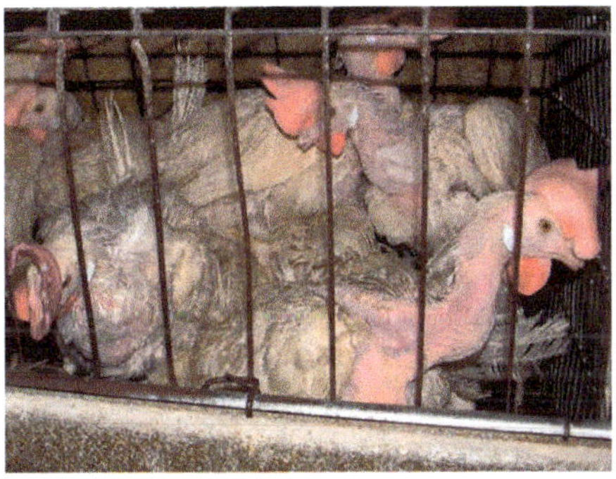

Egg laying hens in a crowded cage.

A chicken egg production facility.

The American Veterinary Medical Association accepts maceration subject to certain conditions, but recommends alternative methods of culling as more humane. Egg-laying hens are then transferred to "battery cageswhere they are kept in high densities. Matheny and Leahy attribute osteoporosis in hens to this caging method. Broiler chicken suffer similar situations, in which they are fed steroids to grow at a super-fast speed, so fast that their bones, heart and lungs often cannot keep up. Broiler chickens under six weeks old suffer painful crippling due to fast growth rates, whilst one in a hundred of these very young birds dies of heart failure.

To reduce aggression in overcrowded conditions, shortly after birth piglets are castrated, their tails are amputated, and their teeth clipped. Calves are sometimes raised in veal crates, which are small stalls that basically immobilize calves during their growth, reducing costs and preventing muscle development, making the resulting meat a pale color, preferred by consumers.

Animal cruelty such as soring, which is illegal, sometimes occurs on farms and ranches, as does lawful but cruel treatment such as livestock branding. Since Ag-gag laws prohibit video or photographic documentation of farm activities, these practices have been documented by secret photography taken by whistleblowers or undercover operatives from such organizations as Mercy for Animals and the Humane Society of the United States posing as employees. Agricultural organizations such as the American Farm Bureau Federation have successfully advocated for laws that tightly restrict secret photography or concealing information from farm employers.

The following are lists of invasive procedures which cause pain, routinely performed on farm animals, and housing conditions that routinely cause animal welfare concerns. It is arguable whether these practices are **cruelty to animals**.

Welfare concerns of farm animals		
Species	**Invasive procedures**	**Housing**
Cattle	• Branding • Castration • Dehorning • Ear tagging • Nose ringing • Restraint • Tail docking • Tongue resection (calves)	• High stocking density (feedlots) • Restricted movement (feedlots) • Veal crates
Pigs	• Castration • Ear docking • Ear tagging • Ear notching • Nose ringing • Tail docking • Tattooing • Teeth cutting • Tusk trimming	• Gestation crates • High stocking density • Restricted movement
Egg laying hens	• Beak-trimming • Blinders • Dubbing	• High stocking density • Restricted movement
Broiler chickens	• Beak-trimming	• High stocking density • Restricted movement

Domestic turkey	• Beak-trimming • Desnooding • Detoeing • Devocalization • Spur removal • Toe clipping	• High stocking density • Restricted movement
Goats and sheep	• Ear tagging • Ear notching • Dehorning • Marking • Mulesing • Tail docking • Teeth grinding	• High stocking density (fine wool industry, live export) • Restricted movement (fine wool industry, live export)

'Blinders' or 'spectacles' are included as some versions require a pin to pierce the nasal septum.

- 'Dubbing' is the procedure of removing the comb, wattles and sometimes earlobes of poultry. Removing the wattles is sometimes called "dewattling.
- 'Desnooding' is the removal of the snood, a fleshy appendage on the forehead of turkeys.
- 'Marking' is the simultaneous mulesing, castration and tail docking of lambs.
- 'Mulesing' is the removal of strips of wool-bearing skin from around the breech (buttocks) of a sheep to prevent flystrike (myiasis)

Psychological Disorders and the link to Human Violence

There are studies providing evidence of a link between animal cruelty and violence towards humans.

Conviction statistics are thought by some to show people convicted for animal cruelty to be more likely to be violent to humans, leading experts to believe that decreasing animal abuse will, in turn, decrease domestic violence. Meanwhile, others explain apparent correlation by criminal courts more often convicting the former for the latter crime as a self-fulfilling prophecy, without any actual link between the two types of actions. Others argue that psychiatry and other authorities outside of courts keep records of who have been cruel to animals and can make biased guesses about whether or not they did violence to humans thereafter and also that they conversely record people who have been violent to humans and can be more biased towards later assuming them to have been cruel to animals, explaining apparent links by institutional bias without link between the actions themselves.

Intentional acts of cruelty can lead to multiple years behind bars. These acts (of intentional animal cruelty or non-accidental injury) may be indicators of serious psychological problems. According to the American Humane Association, 13% of intentional animal abuse cases involve domestic violence. As many as 71% of pet-owning women seeking shelter at safe houses have reported that their partner had threatened and/or actually hurt or killed one or more of their pets; 32% of these women reported that one or more of their children had also hurt or killed pets. Battered women report that they are prevented from leaving their abusers because they fear what will happen to the animals in their absence. Animal abuse is sometimes used as a form of intimidation in domestic disputes.

One of the known warning signs of certain psychopathologies, including antisocial personality disorder, also known as psychopathic personality disorder, is a history of torturing pets and small animals, a behavior known as zoosadism. According to *The New York Times*, "[t]he FBI has found that a history of cruelty to animals is one of the traits that regularly appears in its computer records of serial rapists and murderers, and the standard diagnostic and treatment manual for psychiatric and emotional disorders lists cruelty to animals a diagnostic criterion for conduct disorders.A survey of psychiatric patients who had repeatedly tortured dogs and cats found all of them had high levels of aggression toward people as well, including one patient who had murdered a young boy.Robert K. Ressler, an agent with the Federal Bureau of Investigation's behavioral sciences unit, studied serial killers and noted, Murderers like this (Jeffrey Dahmer) very often start out by killing and torturing animals as kids.

Cruelty to animals is one of the three components of the Macdonald triad, indicators of violent antisocial behavior in children and adolescents. According to the studies used to form this model, cruelty to animals is a common (but not universal) behavior in children and adolescents who grow up to become serial killers and other violent criminals.

It has also been found that children who are cruel to animals have often witnessed or been victims of abuse themselves. In two separate studies cited by the Humane Society of the United States, roughly one-third of families suffering from domestic abuse indicated that at least one child had hurt or killed a pet.

Cultural Rituals

Many times, when Asiatic elephants are captured in Thailand, handlers use a technique known as the training crush, in whichhandlers use sleep-deprivation, hunger, and thirst to ‹break' the elephants' spirit and make them submissive to their owners ; moreover, handlers drive nails into the elephants' ears and feet.

The practice of cruelty to animals for divination purposes is found in ancient cultures, and some modern religions such as Santeria continue to do animal sacrifices for healing and other rituals. Taghairm was performed by ancient Scots to summon devils.

TV and Film Making

Animal cruelty has long been an issue with the art form of filmmaking, with even some big-budget Hollywood films receiving criticism for allegedly harmful—and sometimes lethal—treatment of animals during production.

The American Humane Association (AHA) has been associated with monitoring American filmmaking since after the release of the 1939 film *Jesse James*, in which a horse was pushed off a plank and drowned in a body of water after having fallen 40 feet into it. Initially, monitoring of animal cruelty was a partnership between the AHA and officials in the Hays Office through the Motion Picture Production Code. Provisions in the code discouraged "apparent cruelty to children and animals", and because the Hays Office had the power to enforce this clause, the American Humane Association (AHA) often had access to sets to assess adherence to it. However, because the American Humane Association's Hollywood office depended on the Hays Office for the right to monitor sets, the closure of the Hays Office in 1966 corresponded with an increase in animal cruelty on movie sets.

By 1977, a three-year contract was in place between the Screen Actors Guild (SAG) and the American Federation of Television and Radio Artists which specified that the American Humane Association should be "consulted in the use of animals 'when appropriate'", but the contract did not provide a structure for what "appropriate" meant, and had no enforcement powers. This contract expired in 1980.

One of the most infamous examples of animal cruelty in film was Michael Cimino's legendary flop *Heaven's Gate* (1980), in which numerous animals were brutalized and even killed during production. Cimino allegedly killed chickens and bled horses from the neck to gather samples of their blood to smear on actors for *Heaven's Gate*, and also allegedly had a horse blown up with dynamite while shooting a battle sequence, the shot of which made it into the film. This film played a large part in renewed scrutiny of animal cruelty in films, and led to renewed official on-set jurisdiction to monitor the treatment of animals by the AHA in 1980.

After the release of the film *Reds* (1981), the star and director of the picture, Warren Beatty apologized for his Spanish film crew's use of tripwires on horses while filming a battle scene, when Beatty wasn't present. Tripwires were used against horses when *Rambo III* (1988) and *The 13th Warrior* (1999) were being filmed. An ox was sliced nearly in half during production of *Apocalypse Now* (1979), while a donkey was bled to death for dramatic effect for the Danish film *Manderlay* (2005), in a scene later cut from the film.

There is a case of cruelty to animals in the South Korean film *The Isle* (2000), according to its director Kim Ki-Duk. In the film, a real frog is skinned alive while fish are mutilated. Seven animals were killed for the camera in the controversial Italian film *Cannibal Holocaust* (1980). The images in the film include the slow and graphic beheading and ripping apart of a turtle, a monkey being beheaded and its brains being consumed by natives and a spider being chopped apart. *Cannibal Holocaust* was only one film in a collective of similarly themed movies (cannibal films) that featured unstaged animal cruelty. Their influences were rooted in the films of Mondo filmmakers, which sometimes contained similar content. In several countries, such as the UK, *Cannibal Holocaust* was only allowed for release with most of the animal cruelty edited out.

More recently, the video sharing site YouTube has been criticized for hosting thousands of videos of real life animal cruelty, especially the feeding of one animal to another for the purposes of entertainment and spectacle. Although some of these videos have been flagged as inappropriate by users, YouTube has generally declined to remove them, unlike videos which include copyright infringement.

The Screen Actors Guild (SAG) has contracted with the American Humane Association (AHA) for monitoring of animal use during filming or while on the set.Compliance with this arrangement is voluntary and only applies to films made in the United States. Films monitored by the American Humane Association may bear one of their end-credit messages. Many productions, including those made in the US, do not advise AHA or SAG of animal use in films, so there is no oversight.

Simulations of animal cruelty exist on television, too. On the September 23, 1999 edition of WWE Smackdown!, a plot line had professional wrestler Big Boss Man trick fellow wrestler Al Snow into appearing to eat his pet chihuahua Pepper.

Circuses

The use of animals in the circus has been controversial since animal welfare groups have documented instances of animal cruelty during the training of performing animals. Numerous instances of animal abuse in circuses have been documented such as confining enclosures, lack of regular veterinary care, abusive training methods and lack of oversight by regulating bodies. Animal trainers have argued that some criticism is not based in fact, including beliefs that shouting makes the animals believe the trainer is going to hurt them, that caging is cruel and common, and the harm caused by the use of whips, chains or training implements.

Some circuses now present animal-free acts. Bolivia has enacted what animal rights activists called the world's first ban on all animals in circuses.

Bullfighting

A bull dying in a bullfight.

Bullfighting is criticized by animal rights or animal welfare activists, referring to it as a cruel or barbaric blood sport in which the bull suffers severe stress and a slow, torturous death. A number of activist groups undertake anti-bullfighting actions in Spain and other countries. In Spanish, opposition to bullfighting is referred to as *antitaurismo*.

The Bulletpoint Bullfight warns that bullfighting is "not for the squeamish", advising spectators to "be prepared for blood". It details prolonged and profuse bleeding caused by horse-mounted lancers, the charging by the bull of a blindfolded, armored horse who is "sometimes doped up, and unaware of the proximity of the bull", the placing of barbed darts by banderilleros, followed by the matador's fatal sword thrust. It stresses that these procedures are a normal part of bullfighting and that death is rarely instantaneous. It further warns those attending bullfights tobe prepared to witness various failed attempts at killing the animal before it lies down.

Toro Embolado

The Toro Jubilo, Madrid, 2014.

The "Toro Jubilo" or Toro embolado in Soria, Medinaceli, Spain, is a festival associated with animal cruelty. During this festival, balls of pitch are attached to a bull's horns and set on fire. The bull is then released into the streets and can do nothing but run around in pain, often smashing into walls in an attempt to douse the fire. These fiery balls can burn for hours, and they burn the bull's horns, body, and eyes – all while spectators cheer and run around the victim. The animal rights group PACMA has described the fiesta as "a clear example of animal mistreatment", and PETA calls it "a sadistic festival".

Crush Films

Animal snuff films, known as crush films can be found on the Internet. These films depict instances of animal cruelty, and/or pornographic acts with animals, usually involving the death of an animal, including insects, mice, rats, guinea pigs, hamsters, monkeys, birds, cats, and dogs. In 1999, the U.S. government banned the depiction of animal cruelty; however, the law was overturned by the 3rd U.S. Circuit Court of Appeals which ruled that the category "depiction of animal cruelty" contained in the law was not an exception to First Amendment protections. In an 8–1 decision handed down in April 2010, the U.S. Supreme Court agreed with the lower court's ruling, but on the grounds that the law was unconstitutionally broad. The case itself did not involve crush films, but rather, a video that in part depicted dogfighting.

A horse with a gas mask during World War I.

Warfare

Military animals are creatures that have been employed by humankind for use in warfare. They are a specific application of working animals. Examples include horses, dogs and dolphins. Only recently has the involvement of animals in war been questioned, and practices such as using animals for fighting, as living bombs (as in the use of exploding donkeys) or for military testing purposes (such as during the Bikini atomic experiments) may now be criticised for being cruel.

Princess Anne, the Princess Royal, the patron of the British Animals in War Memorial, stated that animals adapt to what humans want them to do, but that they will not do things that they don't want to, despite training.Animal participation in human conflict was commemorated in the United Kingdom in 2004 with the erection of the Animals in War Memorial in Hyde Park, London.

In 2008 a video of a US Marine throwing a puppy over a cliff during the Iraq conflict was popularised as an internet phenomenon and attracted widespread criticism of the soldier's actions for being an act of cruelty.

Unnecessary Scientific Experiments or Demonstrations

Under all three of the conceptual approaches to animal cruelty discussed above, performing unnecessary experiments or demonstrations upon animals that cause them substantial pain or distress may be viewed as cruelty. Due to changes in ethical standards, this type of cruelty tends to be less common today than it used to be in the past. For example, schoolroom demonstrations of oxygen depletion routinely suffocated birds by placing them under a glass cover, and animals were suffocated in the Cave of Dogs to demonstrate the density and toxicity of carbon dioxide to curious travellers on the Grand Tour.

No pet policies and Abandonment

Many apartment complexes and rental homes institute no pet policies. No pet policies are a leading cause of animal abandonment, which is considered a crime in many jurisdictions. In many cases, abandoned pets have to be euthanized due to the strain they put on animal shelters and rescue groups. Abandoned animals often become feral or contribute to feral populations. In particular, feral dogs can pose a serious threat to pets, children, and livestock. Feral cats are known to hunt endangered birds and have interbred with wild cat populations.

In Ontario, Canada, no pet policies are outlawed under the Ontario Landlord and Tenant Act and are considered invalid even when a tenant signs a lease including a no pets clause. Similar legislation has also been considered in Manitoba.

Laws by Country

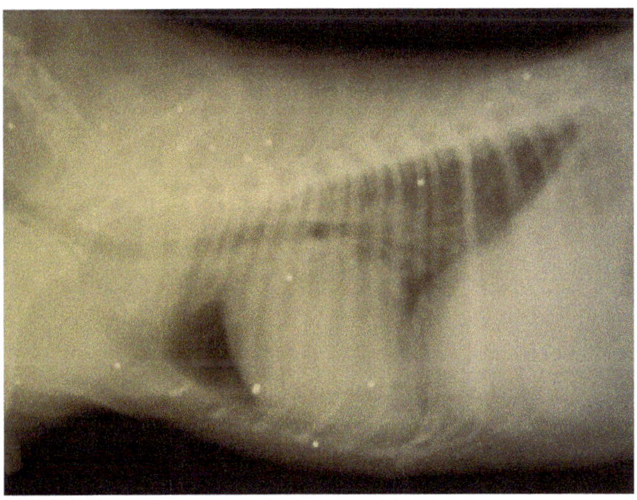

Chest X-ray of a cat that has been shot. White spots are lead shot.

Many jurisdictions around the world have enacted statutes which forbid cruelty to some animals but these vary by country and in some cases by the use or practice.

Africa

Egypt

Egyptian law states that anyone who inhumanely beats or intentionally kills any domesticated animal may be jailed or fined. The Egyptian Society for the Prevention of Cruelty to Animals was established by the British over a hundred years ago, and is currently administered by the Egyptians. The SPCA was instrumental in promoting a 1997 ban on bullfighting in Egypt.

In ancient Egyptian law, the killers of cats or dogs were executed.

South Sudan

The Criminal Code of South Sudan has laws against maltreatment of animals. The laws read:

196. Ill-treatment of Domestic Animal.

Whoever cruelly beats, tortures or otherwise willfully ill-treats any tame, domestic or wild animal, which has previously been deprived of its liberty, or arranges, promotes or organizes fights between cocks, rams, bulls or other domestic animals or encourages such acts, commits an offence, and upon conviction, shall be sentenced to imprisonment for a term not exceeding two months or with a fine.

197. Riding and Neglect of Animal.

Whoever wantonly rides, overdrives or overloads any animal or intentionally drugs or employs any animal, which by reason of age, sickness, wounds or infirmity is not in a condition to work, or neglects any animal in such a manner as to cause it unnecessary suffering, commits an offence, and upon conviction, shall be sentenced to imprisonment for a term not exceeding one month or with a fine or with both.

Americas

Argentina

In Argentina, National Law 14346 sanctions with from 15 days to one year in prison those who mistreat or inflict acts of cruelty on animals.

Brazil

Canada

In Canada, it is an offence under the Criminal Code to intentionally cause unnecessary pain, suffering or injury to an animal. Poisoning animals is specifically prohibited. It is also an offence to threaten to harm an animal belonging to someone else. Most provinces and Territories also have their own animal protection legislation.

The Animal Legal Defense Fund releases an annual report ranking the animal protection laws of every province and territory based on their relative strength and general comprehensiveness. In 2014, the top four jurisdictions were Manitoba, British Columbia, Ontario and Nova Scotia. The worst four were Saskatchewan, Northwest Territories, Quebec, and Nunavut.

Chile

Law 20380 established sanctions including fines, from 2 to 30 Mensual Tributary Units, and prison, from 541 days to 3 years, for those involved in acts of animal cruelty. Also, it promotes animal care through school education, and establishes a Bioethics Committee to define policies related to experiments with animals.

A bull fight in Bogotá, a legacy of Spanish culture. Despite being criticized by numerous organizations in Colombia, the practice of bullfighting remains legal.

Colombia

In Colombia, there is little control over cruel behaviors against animals, and the government has proposed that bullfighting be declared a "Cultural Heritage ; other cruel activities like cockfighting are given the same legal treatment.

Costa Rica

Toucan Grecia with 3D printed prosthetic beak after losing half its beak from a beating by youth.

A new bill has been presented to the Costa Rican legisletive assembly to better define animal cruelty and establish harsher punishments for the abuse of animals. Exceptions for events deemed "traditional", such as horse parades, rodeos and Tico-style bullfights are not provided in the bill's draft which may see motions filed against its passing. The bill had stalled its motion through legislators until notorious toucan Grecia was found having lost half its top beak. The news and images of the injured toucan produced enough contributions to create a 3D printing prosthesis for the bird and rallies in the country's capital for action in relation to the animal cruelty bill.

Mexico

The current policy of Mexico, in civil law, condemns physical harm to animals as property damage to the owners of the abused animal, considering the animals as owned property.

In criminal law, the situation is different. In December 2012, the Legislative Assembly of the Federal District reformed the existing Penal Code of Mexico City, establishing abuse and cruelty to animals as criminal offenses, provided the animals are not deemed to be plagues or pests. Abandoned animals are not considered to be plagues. A subsequent reform was entered into force on January 31, 2013, by a decree published in the Official Gazette of the Federal District. The law provides penalties of 6 months to 2 years imprisonment, and a fine of 50 to 100 days at minimum wage, to persons who cause obvious injury to an animal, and the penalty is increased by one half if those injuries endanger its life. The penalty rises to 2 to 4 years of prison, and a fine of 200 to 400 days at minimum wage, if the person intentionally causes the death of an animal.

This law is considered to extend throughout the rest of the 31 constituent states of the country. In addition, The Law of Animal Protection of the Federal District is wide-ranging, based on banningunnecessary suffering. Similar laws now exist in most states.

United States

The primary federal law relating to animal care and conditions in the US is the Animal Welfare Act of 1966, amended in 1970, 1976, 1985, 1990, 2002 and 2007. It is the only Federal law in the United States that regulates the treatment of animals in research, exhibition, transport, and by dealers. Other laws, policies, and guidelines may include additional species coverage or specifications for animal care and use, but all refer to the Animal Welfare Act as the minimum acceptable standard.

The AWA has been criticized by animal rights groups for excluding birds, rats and mice bred for research, and animals raised for food or fiber as well as all cold-blooded animals.

The Animal Legal Defense Fund releases an annual report ranking the animal protection laws of every state based on their relative strength and general comprehensiveness. In 2013's report, the top five states for their strong anti-cruelty laws were Illinois, Maine, Michigan, Oregon, and California.The five states with the weakest animal cruelty laws in 2013 were Kentucky, Iowa, South Dakota, New Mexico, and Wyoming.

In Massachusetts and New York, agents of humane societies and associations may be appointed as special officers to enforce statutes outlawing animal cruelty.

In 2004, a Florida legislator proposed a ban on "cruelty to bovines," stating: "A person who, for the purpose of practice, entertainment, or sport, intentionally fells, trips, or otherwise causes a cow to fall or lose its balance by means of roping, lassoing, dragging, or otherwise touching the tail of the cow commits a misdemeanor of the first degree." The proposal did not become law.

In the United States, ear cropping, tail docking, rodeo sports, and other acts are legal and sometimes condoned. Penalties for cruelty can be minimal, if pursued. Currently, 46 of the 50 states have enacted felony penalties for certain forms of animal abuse. However, in most jurisdictions, animal cruelty is most commonly charged as a misdemeanor offense. In one recent California case, a felony conviction for animal cruelty could theoretically net a 25-year to life sentence due to their three-strikes law, which increases sentences based on prior felony convictions.

In 2003, West Hollywood, California passed an ordinance banning declawing of house cats. In 2007, Norfolk, Virginia passed legislation only allowing the procedure for medical reasons. However, most jurisdictions allow the procedure.

In April 2013, Texas Federal Court Judge Sim Lake ruled that the Animal Crush Video Prohibition Act of 2010, which criminalized the recording, sale, and transport of videos depicting animal cruelty as obscenity, is in violation of the First Amendment. Judge Lake noted that obscenity tests require an explicitly sexual depiction, which the criminalized videos lack. This follows the precedent set by United States v. Stevens, which additionally held that restrictions on the possession of animal cruelty videos were unconstitutional.

State Welfare Laws

Several states have enacted or considered laws in support of humane farming.

- On November 5, 2002, Florida voters passed Amendment 10 by a margin of 55% for, amending the Florida Constitution to ban the confinement of pregnant pigs in gestation crates.
- On January 14, 2004, the bill AB-732 died in the California Assembly's Agriculture Committee. The bill would have banned gestation and veal crates, eventually being amended to include only veal crates. On May 9, 2007, the bill AB-594 was withdrawn from the California State Assembly. The bill had been effectively killed in the Assembly Agriculture Committee, by replacing the contents of the bill with language concerning tobacco cessation coverage under Medi-Cal. AB-594 was very similar to the current language of Proposition 2.
- On November 7, 2006, Arizona voters passed Proposition 204 with 62% support. The measure prohibits the confinement of calves in veal crates and breeding sows in gestation crates.
- On June 28, 2007, Oregon Governor Ted Kulongoski signed a measure into law prohibiting the confinement of pigs in gestation crates (SB 694, 74th Leg. Assembly, Regular Session).
- In January 2008, Nebraska State Senate bill LB 1148, to ban the use of gestation crates for pig farmers, was withdrawn within 5 days amidst controversy.
- On May 14, 2008, Colorado Governor Bill Ritter signed into law a bill, SB 201, that phases out gestation crates and veal crates.

Venezuela

Venezuela published a "Law for Protection of Domestic Fauna free and in captivity" in 2010, defining responsibilities and sanctions about animal care and ownership. Animal cruelty acts are fined, but are not a cause for imprisonment. The law also forbids the possession, breeding and reproduction of pit bull dogs, among similar breeds that are alleged to be aggressive and dangerous. It elicited reactions from dog owners, who said that aggressiveness in dogs is determined more by treatment by the owner than by the breed itself.

China

Animal welfare and rights in China

As of 2006 there were no laws in China governing acts of cruelty to animals. There are no government supported charitable organizations like the RSPCA, which monitors the cases on animal cruelty. All kinds of animal abuses, such as to fish, tigers, and bears, are to be reported for law enforcement and animal welfare.

Despite the absence of a unified law against animal mistreatment, the World Animal Protection notes that some legislation protecting the welfare of animals exists in certain contexts, especially ones used in research and in zoos.

In September 2009, legislation was drafted to address deliberate cruelty to animals in China. If passed, the legislation would offer some protection to pets, captive wildlife and animals used in laboratories, as well as regulating how farm animals are raised, transported and slaughtered.

In certain jurisdictions such as Fuzhou, dog control officers may kill any unaccompanied dogs on sight. The People's Republic of China is currently in the process of making changes to its stray-dog population laws in the capital city, Beijing. Mr. Zheng Gang who is the director of the Internal and Judicial Committee which comes under the Beijing Municipal People's Congress (BMPC), supports the new draft of the Beijing Municipal Regulation on Dogs from the local government. This new law is due to replace the current Beijing Municipal Regulation on Dog Ownership, introduced in 1889. The current regulation talks ofstrictlylimiting dog ownership and controlling the number of dogs in the city. The new draft focuses instead onstrict management and combining restrictions with management.

Hong Kong

As of 2010, Hong Kong has supplemented or replaced the laws against cruelty with a positive approach using laws that specify how animals should be treated.The government department primarily responsible for animal welfare in Hong Kong is the Agriculture, Fisheries and Conservation Department (AFCD).

Laws enforced by the AFCD include these:

- the Prevention of Cruelty to Animals Ordinance (also enforced by the police)
- the Public Health (Animals and Birds) Ordinance (including regulations for licences imposed on livestock keepers and animal traders and a Code of Standards for Licensed Animal Traders)
- the Dogs and Cats Ordinance
- the Pounds Ordinance
- the Rabies Ordinance
- the Wild Animals Protection Ordinance

In addition, the Food and Environmental Hygiene Department (FEHD) does the following:

- enforces the Public Health and Municipal Services Ordinance, which includes regulations for slaughterhouses and wet markets
- publishes a Code of Practice for the Welfare of Food Animals (which describes their transport)
- publishes Operational Guidelines for the Welfare of Food Animals at Slaughterhouses

The Department of Health does the following:

- enforces the Animals (Control of Experiments) Ordinance.
- publishes a Code of Practice for the Care and Use of Animals for Experimental Purposes

As of 2006, Hong Kong has a law titled "Prevention of Cruelty to Animals Ordinance", with a maximum 3 year imprisonment and fines of HKD$200,000.

India

The Prevention of Cruelty to Animals Act,1960 was amended in the year 1982. According to the newly amended *Indian animal welfare act, 2011* cruelty to animals is an offence and is punishable with a fine which shall not be less than ten thousand Rupees, which may extend to twenty five thousand Rupees or with imprisonment up to two years or both in the case of a first offence. In the case of second or subsequent offence, with a fine which shall not be less than fifty thousand Rupees, but may extend to one lakh Rupees and with imprisonment with a term which shall not be less than one year but may extend to three years.This amendment is currently awaiting ratification from the Government of India. The 1962 Act in the meanwhile is the one that is practiced as of now. The maximum penalty under the 1962 Act is Rs. 50 (under $1). Many organizations, including ones such as the local SPCA, PFA and Fosterdopt are actively involved in assisting the general population in reporting cruelty cases to the police and helping bring the perpetrator to justice. Due to this, much of change has been observed through the subcontinent.

Israel

Japan

In Japan, the 1973 Welfare and Management of Animals Act (amended in 1999 and 2005) stipulates that "no person shall kill, injure, or inflict cruelty to animals without due course", and in particular, criminalises cruelty to all mammals, birds, and reptiles possessed by persons; as well as cattle, horses, goats, sheep, pigs, dogs, cats, pigeons, domestic rabbits, chickens, and domestic ducks regardless of whether they are in captivity.

- Killing or injuring without due reason: up to one year's imprisonment with labour or a fine of up to one million yen
- Cruelty such as causing debilitation by discontinuing feeding or watering without due reason: a fine of up to five hundred thousand yen
- Abandonment: a fine of up to five hundred thousand yen

Separate national and local ordinances exist with regards to ensuring health and safety of animals handled by pet shops and other businesses. Animal experiments are regulated by the 2000 Law for the Humane Treatment and Management of Animals, which was amended in 2006. This law requires those using animals to follow the principles outlined in the 3Rs and use as few animals as possible, and cause minimal distress and suffering. Regulation is at a local level based on national guidelines, but there are no governmental inspections of institutions and no reporting requirement for the numbers of animals used.

Malaysia

Saudi Arabia

Veterinarian Lana Dunn and several Saudi nationals report that there are no laws to protect animals from cruelty since the term is not well-defined within the Saudi legal system. They point to a lack of a governing body to supervise conditions for animals, particularly in pet stores and in the exotic animal trade with East Africa.

South Korea

South Korea's animal welfare laws are weak by international standards.

Taiwan

The Taiwanese Animal Protection Act was passed in 1998, imposing fines up to NT$250,000 for cruelty. Criminal penalties for animal cruelty were enacted in 2007, including a maximum of 1 year imprisonment.

Thailand

Thailand introduced its first animal welfare law in 2014. The *Cruelty Prevention and Welfare of Animal Act, B.E. 2557 (2014)* came into being on 27 December 2014.

Europe

European Union

The European Union Council Directive 1999/74/EC is a directive passed by the European Union on the minimum standards for keeping egg laying hens which effectively bans conventional battery cages. The directive, passed in 1999, banned conventional battery cages in the EU from January 1, 2012 after a 13-year phase-out.

It is also illegal in many parts of Europe to declaw a cat.

France

In France, cruelty to animals is punishable by imprisonment of two years and a financial penalty (30,000 €).

Germany

A German stamp depicting a man beating a chained dog. In the upper left corner it reads: protect the animals.

In Germany, killing animals or causing significant pain (or prolonged or repeated pain) to them is punishable by imprisonment of up to three years or a financial penalty. If the animal is of foreign origin, the act may also be punishable as criminal damage.

Italy

Acts of cruelty against animals can be punished with imprisonment, for a minimum of three months up to a maximum of three years, and with a fine ranging from a minimum of 3,000 Euros to a maximum of 160,000 Euros, as for the law n°189/2004.

Ireland

The *Animal Health and Welfare Act 2013* came into force in 2014, improving animal protection.

Portugal

Since 1 October 2014, violence against animals has been a crime in Portugal. Legislation published in the *Diário da Républica* on August 29 criminalizes the mistreatment of animals, and indicates that "those who, without reasonable cause, inflict pain, suffering, or any other hardship to a companion animal abuse" are to be subject to imprisonment of up to one year. If such acts result in the "death of the animal", the "deprivation of an important organ or member", or "serious and permanent impairment of its capacity of locomotion", those responsible will be punished by imprisonment up to two years.

As for pets, the new law provides thatwhoever, having the duty to store, monitor or pet watch, abandons them, thereby putting in danger their food and the provision of care owedfaces up to six months imprisonment.

Sweden

In Sweden cruelty to animals is punishable by financial penalty and the owner will lose the right to own the animals, and the animals will be removed from the owner.

Switzerland

The Swiss animal protection laws are among the strictest in the world, comprehensively regulating the treatment of animals including the size of rabbit cages, and the amount of exercise that must be provided to dogs.

In the canton of Zurich an animal lawyer, Antoine Goetschel, is employed by the canton government to represent the interests of animals in animal cruelty cases.

Turkey

Under Turkey's Animal Protection Law No. 5199, cruelty to animals is considered a misdemeanor, punishable by a fine only, with no jail time or a black mark on one's criminal record. HAYTAP, the Animal Rights Federation in Turkey, believes that the present law does not contain a strong enough punishment for animal abusers.

United Kingdom

In the United Kingdom, cruelty to animals is a criminal offence for which one may be jailed for up to 6 months.

On August 18, 1911, the House of Commons introduced the Protection of Animals Act 1911 (c.27) following lobbying by the Royal Society for the Prevention of Cruelty to Animals (RSPCA). The maximum punishment was 6 months ofhard labourwith a fine of 25 pounds.

In the London Police Act 1839,fighting or baiting Lions, Bears, Badgers, Cocks, Dogs, or other Animalswas prohibited in London, with a penalty of up to one month imprisonment, with possible hard labour, or up to five pounds. The law laid numerous restrictions on how, when, and where animals could be driven, wagons unloaded, etc.. It also prohibited owners from letting mad dogs run loose and gave police the right to destroy any dog suspected of being rabid or any dog bitten by a suspected rabid dog. The same law prohibited the use of dogs for drawing carts.

Up until then, dogs were used for delivering milk, bread, fish, meat, fruit, vegetables, animal food (the cat's-meat man), and other items for sale and for collecting refuse (the rag-and-bone man). As Nigel Rothfels notes, the prohibition against dogs pulling carts in or near London caused most of the dogs to be killed by their owners as they went from being contributors to the family income to unaffordable expenses. Cart dogs were replaced by people with handcarts. About 150,000 dogs were killed or abandoned. Erica Fudge quotes Hilda Kean:

At the heart of nineteenth-century animal welfare campaigns is the middle-class desire not to be able to see cruelty.

—Hilda Kean, Animal Rights, 1998

The Protection of Animals Act 1911 extended the ban on draft dogs to the rest of the kingdom. As many as 600,000 dogs were killed or abandoned.

The Protection of Animals Act 1911 has since been largely superseded by the Animal Welfare Act 2006, which also superseded and consolidated more than 20 other pieces of legislation, including the Protection of Animals Act 1934 and the Abandonment of Animals Act 1960. The Act introduced the new welfare offence, which means that animal owners have a positive duty of care, and outlaws neglecting to provide for their animals' basic needs, such as access to adequate nutrition and veterinary care.

Under the Criminal Damage Act 1971, domestic animals can be classed as property that is capable of beingdamaged or destroyed. A charge of criminal damage may be appropriate for the injury or death of an animal owned by someone other than the defendant, although prosecution under the Animal Welfare Act 2006 may also be appropriate.

Oceania

Australia

In Australia, all states and territories have enacted legislation governing animal welfare. The legislations are:

- Animal Welfare Act 1992 (ACT)
- Prevention of Cruelty to Animals Act 1979 (NSW)
- Animal Welfare Act (NT)
- Animal Care and Protection Act 2001 (Qld)
- Animal Welfare Act 1985 (SA)
- Animal Welfare Act 1993 (Tas)
- Prevention of Cruelty to Animals Act 1986 (Vic)
- Animal Welfare Act 2002 (WA)

Welfare laws have been criticized as not adequately protecting animals. Whilst police maintain an overall jurisdiction in prosecution of criminal matters, in many states officers of the RSPCA and other animal welfare charities are accorded authority to investigate and prosecute animal cruelty offenses.

REFERENCES

1. "The life of Leonardo da Vinci by Giorgio Vasari". Yale University Library Digital Collections. Retrieved August 21, 2015.
2. Jones, Jonathan (November 30, 2011).Leonardo da Vinci unleashed: the animal rights activist within the artist. The Guardian. Retrieved August 21, 2015.
3. Midgley, Mary (24 May 1999).Descartes' prisoners. Newstatesman.
4. Cassuto, David N. (2007).Bred Meat: The Cultural Foundation of the Factory Farm. Law and Contemporary Problems. 70 *(1): 59–87.*
5. Darwin, Charles (1871). The Descent of Man. p. 34.
6. Norwood, FB; Lusk, JL (2010).Direct versus indirect questioning: An application to the well-being of farm animals. Soc Indic Res. 96 *(3): 551–565.* doi0.1007/s11205-009-9492-z.
7. Are we the cruellest we've ever been? The way we treat animals suggests we are. 19 October 2015. Retrieved 12 November 2015.
8. Pet-Abuse.Com – Animal Cruelty. Pet-abuse.com. Retrieved 2008-11-06.
9. Matheny, Gaverick; Leahy, Cheryl (2007).Farm-Animal Welfare, Legislation, and Trade. Law and Contemporary Problems. 70 *(1): 325–358.*
10. Are Farm Animals Not Considered Animals?. *The Huffington Post. 25 August 2014.*

11. Video: Shocking undercover footage from an egg hatchery – Telegraph. Telegraph. co.uk. 1 September 2009.
12. AVMA Guidelines for the Euthanasia of Animals: 2013 Edition. avma.org
13. Executive Board meets pressing needs.
14. Rev. Sci. Tech. Off. Int. Epiz.,. Global Perspectives on Animal Welfare: Asia, the Far East, and Oceania (n.d.): n. pag. 24 Feb. 2005. Web.
15. Richard A. Oppel, Jr. (April 6, 2013).Taping of Farm Cruelty Is Becoming the Crime. The New York Times. Retrieved April 7, 2013.
16. Sheep dentistry, including tooth trimming. Australian Veterinary Association. Retrieved 1 May 2013.
17. Lockwood, R.; Hodge, G. (1986).The tangled web of animal abuse: The links between cruelty to animals and human violence. Readings in Research and Applications: 77–82.
18. Arluke, A.; Levin, J.; Luke, C.; Ascione, F. (1999).The relationship of animal abuse to violence and other forms of antisocial behavior. Journal of Interpersonal Violence. 14 *(9): 963–975.* doi0.1177/088626099014009004.
19. Alleyne, E., Tilston, L., Parfitt, C. and Butcher, R. (2015).Adult-perpetrated animal abuse: development of a proclivity scale. Psychology, Crime & Law. 21 *(6): 570–588. doi0.1080/1068316X.2014.999064.*
20. Baxendale, S., Lester, L., Johnston, R. and Cross, D. (2015).Risk factors in adolescents' involvement in violent behaviours. Journal of Aggression, Conflict and Peace Research. 7 *(1): 2–18.* doi0.1108/jacpr-09-2013-0025.
21. Animal Cruelty and Human Violence. The Humane Society of The United States. Retrieved 6 October 2015.
22. Delinquent Networks Youth Co-Offending in Stockholm, Jerzy Sarnecki December 2001
23. Hopkins Burke, R. (2001) An Introduction to Criminological Theory, Cullompton
24. Psychiatric Bias and Bigotry in Mainstream Corporate Media, Don Weitz
25. Overt Psychiatric Bias – NICE Guidelines Development Group, 24 October 2006
26. Schram, Jamie (23 March 2013).Brooklyn man furious his roommate wanted to move out allegedly murdered her fish. New York Post. Retrieved 28 August 2014.
27. Green, Kelli (25 March 2013).Man Busted For Killing Fish & Allergic To Exercise. K-FROG 95.1 FM and 92.9 FM. Retrieved 28 August 2014. A roommate dispute resulted in one death, the death of fish. Jose Santiago was charged with animal cruelty and assault after he flushed his roommate's pet fish.
28. Pet-Abuse.Com – Animal Cruelty. Pet-abuse.com. Retrieved 2010-03-17.
29. Loudoun Program Underscores the Link between Domestic Violence, Animal Abuse. Washington Post (September 24, 2014)
30. Facts About Animal Abuse & Domestic Violence Archived November 20, 2012, at the Wayback Machine. American Humane Association Accessed 12 November 2006

31. Domestic Violence & the Animal Abuse Link. Animaltherapy.net. Archived from the original on 2009-12-29. Retrieved 2008-11-06.
32. Felthous, Alan R. (1998). Aggression against Cats, Dogs, and People. In Cruelty to Animals and Interpersonal Violence: Readings in Research and Applications. West Lafayette, Indiana: Purdue University Press. pp. 159–167.
33. Goleman, Daniel (1991-08-07).Clues to a Dark Nurturing Ground for One Serial Killer. New York Times. Retrieved 2010-04-30.
34. Duncan, A.; et al. (2005).Significance of Family Risk Factors in Development of Childhood Animal Cruelty in Adolescent Boys with Conduct Problems. Journal of Family Violence. 20 *(4): 235–239.* doi0.1007/s10896-005-5987-9.
35. Animal Cruelty and Family Violence: Making the Connection. Humane Society of the United States. Archived from the original on October 25, 2008. Retrieved 2008-10-26.
36. Hile, Jennifer (16 October 2002).Activists Denounce Thailand's ElephantCrushingRitual. National Geographic Society. Retrieved 1 October 2014. Just before dawn in the remote highlands of northern Thailand, west of the village Mae Jaem, a four-year-old elephant bellows as seven village men stab nails into her ears and feet. She is tied up and immobilized in a small, wooden cage. Her cries are the only sounds to interrupt the otherwise quiet countryside. The cage is called atraining crush.It's the centerpiece of a centuries-old ritual in northern Thailand designed to domesticate young elephants. In addition to beatings, handlers use sleep-deprivation, hunger, and thirst tobreakthe elephants' spirit and make them submissive to their owners.
37. Turner Classic Movies - TCM.com. Turner Classic Movies. Retrieved 2016-07-20.
38. 27 Jun 1982, Page 14 - The Pantagraph at Newspapers.com. Retrieved 2016-07-20.
39. 30 Sep 1980, Page 56 - St. Louis Post-Dispatch at Newspapers.com. Retrieved 2016-07-20.
40. Andy McKeague, An Interview with Kim Ki-Duk and Suh Jung on The Isle at monstersandcritics.com, May 11, 2005. Retrieved March 11, 2006.
41. Pointless Cannibal Holocaust Sequel in the Works. Fangoria. Retrieved 2007-01-13.
42. Times online, timesonline.co.uk August 19, 2007. Retrieved August 25, 2007.
43. Uproar at fish cruelty on YouTube. practicalfishkeeping.co.uk. May 17, 2007.
44. Entertainment Industry FAQ. Archived from the original on 16 June 2008.
45. Movie Rating System. Earning Our Disclaimer. americanhumane.org
46. Pepper Tribute. Alsnowshead.tripod.com (1999-09-03). Retrieved on 2011-12-14.
47. The Wrestling Menu #25 – The History Of Al Snow. December 4, 2002
48. Circus Incidents: Attacks, Abuse and Property Damage(PDF). Humane Society of the United States. 2004-06-01. Retrieved 2015-12-15.
49. Circuses. Humane Society of the United States. Retrieved 2015-12-15.

50. Patton, K (2007-04-01).Frequently Asked Questions: Do circus trainers/handlers abuse animals?. lionden.com. Retrieved 2008-05-23.
51. Circuses. Peta.org. Retrieved 2015-12-15.
52. Animal free circuses(PDF).
53. Bolivia bans all circus animals. Associated Press (via Guardian). 31 July 2009. Retrieved on 2011-12-14.
54. What is bullfighting?. League Against Cruel Sports. Archived from the original on September 30, 2011.
55. Running of the Bulls Factsheet.
56. ICABS calls on Vodafone to drop bullfighting from ad. Irish Council Against Blood Sports.
57. The suffering of bullfighting bulls.
58. The Bulletpoint Bullfight, p. 6, ISBN 978-1-4116-7400-4
59. Typically Spanish – Spain News, Toro Jubilo fiesta returns to Medinaceli, Soria
60. PETA, Help Stop the Sadistic 'Fire Bull' Festival in Soria
61. 'Toro Jubilo' 2011 en Medinaceli, on YouTube
62. 3rd Circuit Strikes Down Law Criminalizing Sale of Animal Cruelty Depictions. Law.com (2008-07-22). Retrieved on 2011-12-14.
63. Sherman, Mark (April 10, 2010).Court voids law aimed at animal cruelty videos. Associated Press. Retrieved April 10, 2010.
64. Animals in War – The unseen casualties. Animal Aid. 2003-06-01. Retrieved 2009-09-05.
65. The Military's War on Animals. PETA. Retrieved 2009-09-05.
66. Price, Eluned (2004-11-01).They served and suffered for us. London: The Daily Telegraph. Retrieved 2009-09-05.
67. Animal war heroes statue unveiled. The Daily Telegraph. 2004-11-24. Retrieved 2009-09-05.
68. Naughton, Philippe (4 March 2008).Puppy-toss video makes Marine figure of hate. London: The Times. Retrieved 4 September 2009.
69. L'air et le monde aérien.
70. Taylor, An (1832).Account of the Grotta del Cane; With Remarks Upon Suffocation by Carbonic Acid. The London Medical and Physical Journal: 278–285.
71. Fleming & Johnson, Toxic Airs: Body, Place, Planet in Historical Perspective, Pittsburgh, 255–256.
72. Kroonenberg, Why Hell Stinks of Sulfur: Mythology and Geology of the Underworld, Chicago, 2013, 41–45.
73. U.S. Facing Feral-Dog Crisis.
74. Why no-pet rental clauses lack teeth. thestar.com. 7 December 2012.

75. No-pet policy for Man. renters could be outlawed. 15 February 2010.
76. Legislature Related to Animals in Egyptian Law
77. Humanity, through animal care. Weekly.ahram.org.eg (2003-09-10). Retrieved on 2011-12-14.
78. (Not-So-) BIZARRE DOG LAW California Man Faces Life in Prison for Killing Dog; and Tennessee Judge Slam-Dunks Puppy Mill Owners July 14, 2002 Dogs in the News Archived December 19, 2009, at the Wayback Machine.
79. Dennis C. Turner (26 June 2000). The domestic cat: the biology of its behaviour. Cambridge University Press. pp. 185–. ISBN 978-0-521-63648-3. Retrieved 15 December 2011.
80. The Penal Code Act, 2008. sudantribune.com
81. LEY 14346 – MALOS TRATOS Y ACTOS DE CRUELDAD A LOS ANIMALES (PDF). Gobierno República de Argentina. Retrieved 7 September 2013.
82. Cruelty to Animals , Criminal Code, s. 445.1.
83. "Cattle and Other Animals", Criminal Code, s. 445.
84. "Assaults", Criminal Code, s. 264.1(1)(c).
85. "A Report on Animal Welfare Law in Canada", Alberta Farm Animal Care, June 2004.
86. "2014 Canadian Animal Protection Laws Rankings", June 5, 2014, Animal Legal Defense Fund, report available for download at link.
87. Ley 20380, SOBRE PROTECCIÓN DE ANIMALES(PDF) (in Spanish). MINISTERIO DE SALUD; SUBSECRETARÍA DE SALUD PÚBLICA. October 2009. Retrieved 7 September 2013.
88. Colombian president offers to grant bullfighting status of cultural heritage. Retrieved 17 October 2012.
89. Aias, L (4 March 2016).Costa Rica moves to set prison sentences for animal cruelty. The Tico Times. Retrieved 14 September 2016.
90. Khan, Carrie (27 Aug 2016).After Losing Half A Beak, Grecia The Toucan Becomes A Symbol Against Abuse. NPR. Retrieved 14 Sep 2016.
91. Finally, reform to the Penal Code of Mexico City on animal abuse. Retrieved 9 February 2013.
92. Lawmakers seek to position Federal enforcement of the Law of Animal Protection of the Federal District. Retrieved 9 February 2013.
93. Legislative History of the Animal Welfare Act. Archived from the original on 2 March 2013. Retrieved 7 April 2010.
94. The Animal Welfare Act. Archived from the original on April 29, 2009. Retrieved 7 April 2010.
95. Annual Study Names 2013'sTop Five States to be an Animal Abuser. Animal Legal Defense Fund.
96. Book Review: Brute Force: Animal Police and the Challenge of Cruelty. Ccja-

acjp.ca. Retrieved on 2011-12-14.

97. Emery, David.Florida to Consider Ban on Cow Tipping. About.com. Retrieved 2007-06-07.
98. ALDF: U.S. Jurisdictions With and Without Felony Animal Cruelty Provisions. Aldf.org. Retrieved 2009-04-29.
99. Accused Dog Killer Could Get 25 Years to Life in Prison. Retrieved 2008-11-06.
100. Judge allows California cities to ban cat declawing. Sfgate.com (2007-10-11). Retrieved on 2011-12-14. Archived December 18, 2007, at the Wayback Machine.
101. Norfolk Bans De-Clawing Of Cats Archived December 11, 2008, at the Wayback Machine.
102. 'Animal crush' video charges dismissed in first case under new law. POLITICO.
103. PorkNet Newsletter. MetaFarms.com, Inc. 2002-11-07. Archived from the original on 2006-03-14. Retrieved 2008-07-03.
104. Criminal Justice and Judiciary. California State Senate. 2004. Archived from the original on February 18, 2011.
105. AB-732 Analysis. California State Assembly. 2008-01-14.
106. 2007 Mid Year Summary. California Assembly Committee on Agriculture. 2007. Archived from the original on January 10, 2010.
107. AB-594 Analysis. California State Assembly. 2008-05-09.
108. Andrea Johnson, "Polls Indicate Strong Support for Pen Gestation for Hogs. March 29, 2007
109. Back door activists gain momentum. Learfield Communications, Inc. 2007-07-05. Retrieved 2008-07-03.
110. Farm Animal Welfare Bill Killed in Legislature. Omaha World Daily. 2008-02-17.
111. Farm Sanctuary Applauds Colorado for Passing Legislation Phasing out Veal and Gestation Crates. Reuters. 2008-05-14. Retrieved 2008-07-03.
112. Farm Animal Welfare Measure Becomes Law. Federation of Animal Science Societies (FASS). 2008-05-14. Retrieved 2008-07-03.
113. Asamblea Nacional de la República Bolivariana de Venezuela.LEY PARA LA PROTECCIÓN DE LA FAUNA DOMÉSTICA LIBRE Y EN CAUTIVERIO(PDF). Diario El Universal. Retrieved 7 September 2013.
114. Joaquín Perez Guisado, Andre Muñoz Serrano (2009).Factors Linked to Dominance Aggression in Dogs(PDF). Journal of Animal and Veterinary Advances.
115. Richard Spencer. Just who is the glamorous kitten killer of Hangzhou? April 3, 2006.
116. SBS Australia.The Biggest Chinese Restaurant in the World. Retrieved 4 November 2008
117. Journal of Ecotourism.The Shark Watching Industry and its Potential Contribution to Shark Conservation. Retrieved 8 November 2008.

118. Sohu Forum.人類的飲食與野生動物的滅絕有著本質和必然的聯繫". Archived from the original on 23 March 2012. Retrieved 8 November 2008.

119. 中國青年報. "國家禁令擋不住虎骨酒熱銷". Retrieved 4 November 2008.

120. Jadecampus.Conservationists Call on China to Support Law Over Tiger Farms. Retrieved 4 November 2008.

121. 中國青年報. "拿什麼拯救你可憐的黑熊:能不能不用熊膽?". Retrieved 4 November 2008.

122. China. World Animal Protection. Retrieved 23 February 2016.

123. China unveils first ever animal cruelty legislation. The Daily Telegraph. London. September 18, 2009. Retrieved 2009-09-18.

124. Beijing loosens leash on pet dogs. Chinadaily.com.cn. Retrieved 2008-11-06.

125. Review of Animal Welfare Legislation in Hong Kong by Amanda S. Whitfort and Fiona M. Woodhouse, June 2010. This document reviews animal welfare laws and compares them to those of Taiwan, Singapore, Australia, New Zealand, the European Union, and the United States.

126. Agriculture, Fisheries and Conservation Department, "Penalty for Cruelty to Animals," Prevention of Cruelty to Animals (Chapter 169, Section 3) 15 December 2006

127. The prevention of cruelty to animals act,1960(PDF). Amendments. Ministry of environment and Forests,Government of India. Retrieved 9 February 2012.

128. Indian animal welfare act 2011(PDF). Chapter IV.Cruelty to animals. Ministry of Environment and Forests,Government of India. Retrieved 9 February 2012.

129. THE ANIMAL WELFARE ACT, 2011. awbi.org

130. Act on Welfare and Management of Animals (Act No. 105 of October 1, 1973). (PDF). cas.go.jp (in Japanese and English). Retrieved on 2011-12-14.

131. Christopher S. Stevenson, Lisa A. Marshall and Douglas W. Morgan Japanese guidelines and regulations for scientific and ethical animal experimentation. Progress in Inflammation Research 2nd Edition 2006 p. 187. doi0.1007/978-3-7643-7520-1_10

132. Select Committee on Animals In Scientific Procedures Report July 2002, Accessed 23rd August 2007

133. Animal lovers lament lack of law against cruelty. Arabnews.com (2009-03-12). Retrieved on 2011-12-14. Archived September 9, 2015, at the Wayback Machine.

134. World Animal Protection (November 2, 2014).Korea. Retrieved May 8, 2016.

135. Koahsiung Municipal Institute for Animal Health,Laws and Regulations,Animal Protection Act last amended 11 July 2007

136. CRUELTY PREVENTION AND WELFARE OF ANIMAL ACT, B.E. 2557 (2014). Thai SPCA. Archived from the original on 13 May 2016. Retrieved 11 May 2016.

137. Kanchanalak, Pornpimol (13 November 2014).A landmark victory for animal rights. The Nation. Retrieved 22 November 2015.

138. COUNCIL DIRECTIVE 1999/74/EC of 19 July 1999. European Union
139. Declawing Cats: Manicure or Mutilation?. dehumane.org
140. Code pénal. legifrance.gouv.fr. Retrieved November 28, 2012.
141. Tierschutzgesetz (TierSchG). Bundesrecht.juris.de. Retrieved on 2011-12-14.
142. Strafgesetzbuch (StGB). Bundesrecht.juris.de. Retrieved on 2011-12-14.
143. The Italian Parliament – Law 189/2004 – Art. 544/ter/quater/quinquies. Camera.it. Retrieved on 2011-12-14. (in Italian)
144. Animal Health and Welfare Act 2013.
145. New animal welfare legislation introduced. RTE.ie. 7 March 2014.
146. The Agency.Boas Notícias – Animais: Lei que criminaliza maus-tratos entra em vigor.
147. Brottsbalk (196200).
148. Scales of Justice: In Zurich, Even Fish Have a Lawyer, Deborah Ball. The Wall Street Journal. March 6, 2010
149. The lawyer who defends animals, Leo Hickman. The Guardian. March 5, 2010
150. ANIMAL PROTECTION BILL LAW no 5199, HAYTAP, accessed December 7, 2012
151. "Civil society skeptical about amendment to animal protection law", HAYTAP, accessed December 7, 2012
152. "HAYTAP : Animal Rights Federation in Turkey", HAYTAP, accessed December 7, 2012
153. Animal Welfare Act 2006. Chapter 45. (PDF). opsi.gov.uk. Retrieved on 2011-12-14.
154. The Times, Monday, Jan 01, 1912; p. 3; Issue 39783; col F "The Animals' New Magna Charter"
155. London Police Act 1839, Great Britain Parliament. Section XXXI, XXXIV, XXXV, XLII. Retrieved 2011-01-23.
156. /Graham Robb (2007). The discovery of France: a historical geography from the Revolution to the First World War. W. W. Norton & Company. pp. 167–. ISBN 978-0-393-05973-1. Retrieved 15 December 2011.
157. Dog Carts and the Extinction of Memory. October 15, 2008
158. Rothfels, Nigel, Representing Animals, Indiana University Press, p. 12, ISBN 0-253-34154-X. Chapter: ‹A Left-handed Blow: Writing the History of Animals' by Erica Fudge
159. igg.org.uk.
160. Animal Rights by Hilda Kean, 1998, University of Chicago Press. ArchivedNovember 11, 2010, at the Wayback Machine.
161. Protection of Animals Act 1911. Animallaw.info (1911-08-18). Retrieved on 2011-12-14.

162. Pet abuse law shake-up unveiled. BBC News Online. 14 October 2005.
163. BBC – Ethics – Animal Ethics: Animal Welfare Act. BBC. Retrieved 24 April 2010.
164. Cats And The Law – Cats Away. Retrieved 13 January 2015.
165. Offences involving Domestic and Captive Animals. The Crown Prosecution Service. Retrieved 13 January 2015.
166. What is the Australian legislation governing animal welfare?.
167. ACT legislation register – Animal Welfare Act 1992 – main page.
168. ANIMAL WELFARE ACT 1992.
169. NSW Legislation.
170. PREVENTION OF CRUELTY TO ANIMALS ACT 1979.
171. ANIMAL WELFARE ACT.
172. ANIMAL WELFARE ACT.
173. Animal Care and Protection Act 2001 (Qld)(PDF). legislation.qld.gov.au.
174. ANIMAL CARE AND PROTECTION ACT 2001.
175. South Australian Legislation.
176. ANIMAL WELFARE ACT 1985.
177. Legislation View Page.
178. ANIMAL WELFARE ACT 1993.
179. Prevention of Cruelty to Animals Act 1986 (Vic)(PDF). legislation.vic.gov.au.
180. PREVENTION OF CRUELTY TO ANIMALS ACT 1986.
181. Western Australian Legislation.
182. ANIMAL WELFARE ACT 2002.
183. Graeme McEwen. The fox is in charge of the chickens Animals Australia. Retrieved July 4, 2008.
184. Animal Welfare Act 1999.

Index